AF718408

The Shortcut Guide to

Achieving Business Intelligence in Midsize Companies

By Don Jones

Realtime Publishers – San Francisco, CA

The Shortcut Guide to Achieving Business Intelligence in Midsize Companies

Published by:
Realtime Publishers
300 Montgomery Street
Suite #1121
San Francisco, CA 94104
www.realtimepublishers.com

Copyright © 2010 by Realtime Publishers, San Francisco, California

Copyright Statement

© 2010 Realtime Publishers. All rights reserved. This site contains materials that have been created, developed, or commissioned by, and published with the permission of, Realtime Publishers (the "Materials") and this site and any such Materials are protected by international copyright and trademark laws.

THE MATERIALS ARE PROVIDED "AS IS" WITHOUT WARRANTY OF ANY KIND, EITHER EXPRESS OR IMPLIED, INCLUDING BUT NOT LIMITED TO, THE IMPLIED WARRANTIES OF MERCHANTABILITY, FITNESS FOR A PARTICULAR PURPOSE, TITLE AND NON-INFRINGEMENT. The Materials are subject to change without notice and do not represent a commitment on the part of Realtime Publishers its web site sponsors. In no event shall Realtime Publishers or its web site sponsors be held liable for technical or editorial errors or omissions contained in the Materials, including without limitation, for any direct, indirect, incidental, special, exemplary or consequential damages whatsoever resulting from the use of any information contained in the Materials.

The Materials (including but not limited to the text, images, audio, and/or video) may not be copied, reproduced, republished, uploaded, posted, transmitted, or distributed in any way, in whole or in part, except that one copy may be downloaded for your personal, non-commercial use on a single computer. In connection with such use, you may not modify or obscure any copyright or other proprietary notice.

The Materials may contain trademarks, services marks and logos that are the property of third parties. You are not permitted to use these trademarks, services marks or logos without prior written consent of such third parties.

Realtime Publishers and the Realtime Publishers logo are registered in the US Patent & Trademark Office. All other product or service names are the property of their respective owners.

If you have any questions about these terms, or if you would like information about licensing materials from Realtime Publishers, please contact us via e-mail at info@realtimepublishers.com.

For questions about Realtime Publishers, visit: http://www.realtimepublishers.com

To download any Realtime books, visit our library: http://nexus.realtimepublishers.com

ISBN: 978-1-935581-11-6

Manufactured in the United States of America

About the Author

Don Jones has more than a decade of professional experience in the IT industry. He's the author of more than 30 IT books, including *Windows PowerShell: TFM; VBScript, WMI, and ADSI Unleashed; Managing Windows with VBScript and WMI*; and many more. He's a top-rated and in-demand speaker at conferences such as Microsoft TechEd and TechMentor, and writes the monthly Windows PowerShell column for Microsoft TechNet Magazine. Don is a multiple-year recipient of Microsoft's "Most Valuable Professional" (MVP) Award with a specialization in Windows PowerShell. Don's broad IT experience includes work in the financial, telecommunications, software, manufacturing, consulting, training, and retail industries and he's one of the rare IT professionals who can not only "cross the line" between administration and software development, but also between IT workers and IT management. Don is a co-founder of Concentrated Technologies, and serves as author and series editor for Realtime Publishers.

Table of Contents

Chapter 1: What Is Business Intelligence, Anyway? 1

Business Goals for BI 2

BI: Not Just for Big Companies 3

BI Terminology 4

Data Warehouses and Data Marts 4

Data Mining 6

Reporting, Dashboards, and Scorecards 6

Predictive Analytics 9

Business Performance Management 10

A Brief Technical Overview of BI 11

Data Marts and Data Warehouses 11

OLTP vs. OLAP 11

Normalized vs. Dimensional Data 12

Data Modeling 14

Data Transformation and Loading 14

In-Memory Analytics 15

Reporting, Analysis, and Visualization 15

BI: Analysis + Planning 18

Analyzing and Reporting on Information 18

Making Decisions 19

Implementing 20

One Size Does Not Fit All 20

BI in Large Enterprises....20

BI in Midsize Companies....21

Coming Up Next....21

Chapter 2: How Business Intelligence Happens....23

Assembling Your Business Data....23

Direct Data Mapping....24

Data Loading....25

Data Marts and Data Warehouses....26

In-Memory Analytics....28

Creating an Information-Driven Business Culture....30

Reporting....31

Dashboards....33

Scorecards....34

Analyzing Business Information....36

Analysis and Visualization....37

Uncovering Root Causes Behind Data....42

The Importance of Modeling the Business....43

Making Decisions and Taking Action—The Planning Connection....44

Planning: Aligning Resources to Opportunities....45

Coming Up Next....48

Chapter 3: Debunking the Top-Three Myths of Business Intelligence for Midsize Companies....49

Myth 1: BI Can Be Disruptive....49

Specialized Tools and Training Required....50

Long Implementation Times 53

Impact on Other Business Processes 53

Myth 2: BI Requires Specialized Expertise 54

Complicated Deployment and Implementation 55

Specialized Technology Management Skills 55

Large-Company Skills Don't Scale Down 56

Specialized End-User Skills 57

Myth 3: BI Is Expensive 60

Expensive Acquisition Costs 60

Expensive Consulting Services 61

Don't Like Negotiating with Vendors? 61

Unique Advantages of Midsize Business BI 62

In-Memory Analytics 63

Consistency 63

What's Good for Big Business Is Kind of Good for You Too 63

Business Model Flexibility 64

Advanced Reporting 64

Data-Everywhere Accessibility 65

Coming Up Next 67

Chapter 4: Successfully Adding Business Intelligence to a Midsize Company 68

Reviewing the Problems with BI 68

Too Complex 68

Too Expensive 70

Too Disruptive ... 70

Adding BI Without Breaking the Bank ... 71

Not Just Reporting! ... 71

Think Big, Start Small ... 72

Start with the Biggest Pain ... 73

Add Capabilities as Needed ... 75

Ensure that You Have a Growth Path ... 77

Seek Out Fixed-Price Solutions ... 77

Adding BI with Your Current IT Team ... 78

Software Appliances ... 78

Modular Appliances ... 79

Ensure that You Have a Growth Path ... 80

Midsize Manageability ... 80

Adding BI without Disrupting Business ... 82

Low Startup Cost ... 82

Minimal Specialized Expertise ... 82

Self-Service via Web Consoles ... 82

Intuitive User Interfaces ... 83

Keep Your Existing Tools ... 85

What's Good for Big Business...Isn't Necessarily Good for You ... 85

Download eBooks from Realtime Nexus! ... 88

Chapter 1: What Is Business Intelligence, Anyway?

For a few years now, *business intelligence* have been buzzwords of the IT industry. Unlike a lot of IT buzzwords—like "Web 2.0"—business intelligence actually carries some weight, has a definite meaning, and brings real value to businesses. Business intelligence—or BI, as insiders like to call it—refers to the practice of helping a business acquire a better understanding of itself. More broadly, BI also refers to the skills, technologies, applications, and practices involved in bringing that understanding to light.

The term *business intelligence* was first used in a 1958 article by IBM researcher Hans Peter Luhn, who defined it as *the ability to apprehend the interrelationships of presented facts in such a way as to guide action towards a desired goal.* In other words, BI isn't just about technology-centric concepts such as data warehousing or business analytics; BI is really about understanding the relationships between different aspects of your company so that you can guide the company toward specific goals, like increasing market share and improving customer satisfaction.

The IT industry—being technology-centric, of course—uses the term *BI* as a sort of umbrella phrase that covers all the technologies and capabilities used to gather facts about the business, present those facts in a way that makes relationships clearer, and allow manipulation of those facts to project "what if" scenarios—all intended to help guide better decision making.

Let's consider a simple example of how BI can help a business.

Improving Customer Satisfaction

Widgets, Inc. has been struggling to improve their customer service levels. The company generally gets good customer satisfaction scores on the surveys it conducts, but it goes through periods where satisfaction drops by more than 30%. The company's executives already know that the drops come during periods when the company is extremely busy, fielding far more orders than usual. The company has already spent tens of thousands of dollars improving their distribution center operations to reduce the time it takes to ship customer orders, but it hasn't seemed to make any difference.

The company invests in a BI system. The system gathers information from a number of internal sources, including the main order-processing database. The system also collects data from some external sources, including billing data from the company's shipping vendors, the company's payroll system, and other places.

After using the BI system for a few months, company managers notice something unexpected. During periods of decreased customer satisfaction, their distribution center is actually *less* busy—they can see the increase in customer orders correlate with a *decrease* in payroll for the distribution center. Digging a bit further, they realize that the increased customer orders are mainly for products that are being drop-shipped directly from a couple of specific vendors to customers. Management realizes that it's those *vendors* that are slow to fulfill orders, causing the drop in customer satisfaction. They now know to focus their efforts on improving those vendors' performance, finding new vendors for those products, or stocking those products in their own distribution center, where they'll have better control over shipping times.

The driving force behind BI is that companies are drowning in unrelated facts that come from *silos*: payroll data, financial data, customer data, vendor data, and so on. BI pulls all that data together and correlates it. The data may seem unrelated, but in fact *everything* in the business is related somehow—if some data is *truly* unrelated, then why is it in the business in the first place? BI doesn't generate new data—it simply makes it easier to explore overlooked relationships between data.

Business Goals for BI

Most companies—even midsize ones—have an incredible amount of data living in transaction-based, distributed systems and databases. The payroll system has one database, the order-processing system has another, and so on. These databases are typically fine tuned for individual transactions, such as retrieving a single customer order, or for specific batch operations, such as processing payroll at the end of each month. What these databases are *not* designed to do is communicate with one another, to allow users to explore data in unusual ways, or to provide high-level summaries of the data in an instant. The main goal for BI, then, is to provide exactly those things:

- The ability to view data from multiple sources in a single view, such as seeing sales information correlated with shipping expenses and facility utility bills
- The ability to quickly see summaries of data from different places—such as the total payroll spent, along with the total sales, for a given time period
- The ability to see data over time, comparing data from yesterday, last month, the past three quarters of the past 5 years, to see how things have changed over time
- The ability to ask "what if" questions and have answers generated based on historical data; for example, you might want to know if raising sales by 10% for a sustained period of time will necessitate a raise in utility costs or payroll expenses

Ideally, the answers to these questions can feed directly into the company's planning systems, helping define budgets, sales goals, and other planning elements. Doing so allows historical trends to drive business decisions, and those decisions automatically drive business planning.

One of the reasons BI has become so popular a term in the past few years is that it is an *incremental* IT investment. You don't have to tear apart your existing systems in any way—in fact, those systems won't know that they're participating in a BI solution. The fact that BI doesn't impact your production systems means implementing BI is relatively low risk—you're not likely to disrupt daily operations while implementing a BI solution. Best yet, a properly-designed and –implemented BI solution can quickly deliver a high return on investment (ROI), something executives really appreciate.

BI: Not Just for Big Companies

There's a perception that BI is just for major enterprises, and it's certainly true that major enterprises have a lot to gain from implementing BI and that those big companies have the extra cash needed to implement a BI solution. Implementation in major enterprises can take months, and it's rarely an inexpensive undertaking.

But that doesn't mean midsize companies can't benefit from BI, and it doesn't mean midsize companies have to spend as much time or money as huge companies do. After all, midsize companies often use less-expensive, easier-to-implement solutions for things like payroll, corporate bookkeeping, and customer relationship management; BI doesn't need to be any different.

But there's an important corollary to that statement: Just as midsize businesses *don't* use the same bookkeeping software that a major enterprise does, a midsize business *won't* use the same BI solution that a gigantic company uses. Bringing a large-scale BI solution into a midsize company isn't any smarter than bringing a gigantic facilities-management package into a midsize company. What midsize companies need is a midsize BI package. Huge companies are generally accustomed to lengthy implementation times for *any* new solution package—whether it's Enterprise Resource Planning (ERP), Customer Relationship Management (CRM), payroll, or whatever. Midsize companies typically take advantage of more prepackaged solutions that are less expensive and much easier to implement. The same approach can work for BI.

In fact, you can make an argument that midsize companies can see a quicker, better return on a BI investment than large companies. After all, large companies can be slow to change, even when they've got good intelligence, smart decisions, and a good plan. Even with a fully-implemented BI system, large companies often take longer to implement the decisions that their BI leads them to. Midsize companies, in contrast, are known for being somewhat simpler in structure and can often react more quickly to change. With the right BI solution, a midsize company could easily develop solid BI, make smart decisions about the company's future, and *act* on those decisions more rapidly—meaning a faster ROI from that BI solution.

BI Terminology

Let's briefly step back from the business aspects of BI and look at some of the major elements that go into a BI solution. We'll start with the next five sections, where I'll outline the major aspects of a BI solution at a high level.

Data Warehouses and Data Marts

Central to most BI efforts is a *data mart* or a *data warehouse.* They're both basically the same thing under the hood: a specialized kind of database that's designed to support business analytics and to contain data from one or many different sources. The difference between a mart and a warehouse is their scope: A data *mart* only seeks to serve the needs of a portion of the company, such as the marketing department or finance department. A data *warehouse* seeks to serve the entire company.

There are two design approaches that come into play when you're talking about marts and warehouses. The *top-down* approach directs us to build the data warehouse first, considering the needs of the entire company when designing it. That's obviously a complex task, as you really need to look at how the entire company works and how its various pieces fit together. With the warehouse in place, you can create data marts that extract a portion of the warehouse to meet the needs of specific audiences, such as finance or marketing.

The *bottom-up* approach directs the opposite, having us build data marts for each audience within the company, then combining those to form a companywide data warehouse. This is often easier to implement, simply because it's easier to wrap your brain around the needs of a specific department—such as marketing—than to grasp the companywide view of things. However, this approach is not without its detractors. In a 2000 article, "Data Mart Does Not Equal Data Warehouse," author William Inmon offered this analogy: "You can catch all the minnows in the ocean and stack them together and they still do not make a whale." What he's saying is that the silo-specific data marts will have gaps between them, and they'll never, in sum, provide the real companywide view you need.

These sorts of design issues are what make BI implementation in large companies so complex. In midsize companies, however, there's an advantage: Prepackaged solutions can take much of this controversy out of your hands by providing a pre-made data warehouse.

Ralph Kimball, another noted data warehouse design expert, favors what many describe as a bottom-up approach. He suggests that, first and foremost, you need to focus on your business. Building smaller data marts that each focus on a particular subject within your business helps constrain the task to something more manageable, and helps you keep the business—not the ultimate data warehouse—more firmly in mind. He recommends building data marts not around business units but around business processes, such as orders, shipments, payments, and so forth. Kimball refutes the use of the "bottom-up" label for his approach, pointing out that his approach doesn't follow the traditional bottom-up approach of designing around organizational units within the business but instead around how the business works and what it does.

Resource
A full discussion of Ralph Kimball's approach is beyond the scope of this book, but you can learn more at http://www.ralphkimball.com/. Especially informative is his "Fables and Facts" page at http://www.ralphkimball.com/html/controversies.html.

So what *is* a data warehouse? I'll get into the technical details a bit later in this chapter, but for now, think of a data warehouse as a place where *all* your company's data is *copied to* and then rearranged so that relationships are easier to perceive and summaries are easier to generate.

Data Mining

Data mining is simply the process of extracting patterns from data—in BI terms, from a data warehouse. Want to find out why the distribution center processes orders more slowly at certain times of the month? Mine your data warehouse looking for patterns—perhaps you'll find that things slow down as the supply of cardboard boxes dwindles because the box vendor isn't fulfilling *their* orders quickly enough.

It's important to realize, however, that data mining can't reveal patterns in data that aren't present in the data being mined. That sounds obvious, but it can be deceptive because data mining can often *seem* to reveal patterns that aren't really there. What you can find yourself looking at is a pattern comprised of *symptoms* rather than *causes,* when the causal data isn't present in the data warehouse. This isn't to say that data mining isn't useful—it's at the heart of BI, in fact. Rather, you just need to be aware of what's in the data, and what isn't, and take common-sense approaches to verifying and validating the conclusions to which your data mining leads you.

Reporting, Dashboards, and Scorecards

With all the data that a data warehouse can contain, most users will need simplified means of looking at commonly-examined information. Reports are one obvious product of a data warehouse, and they can range from high-level summaries to extremely detailed analyses. Another option is a *dashboard,* which provides a summary of common metrics possibly from multiple sources, and often contains planning and actual comparisons, often visualized in a slick, simplified user interface (UI)—such as the example Figure 1.1 shows.

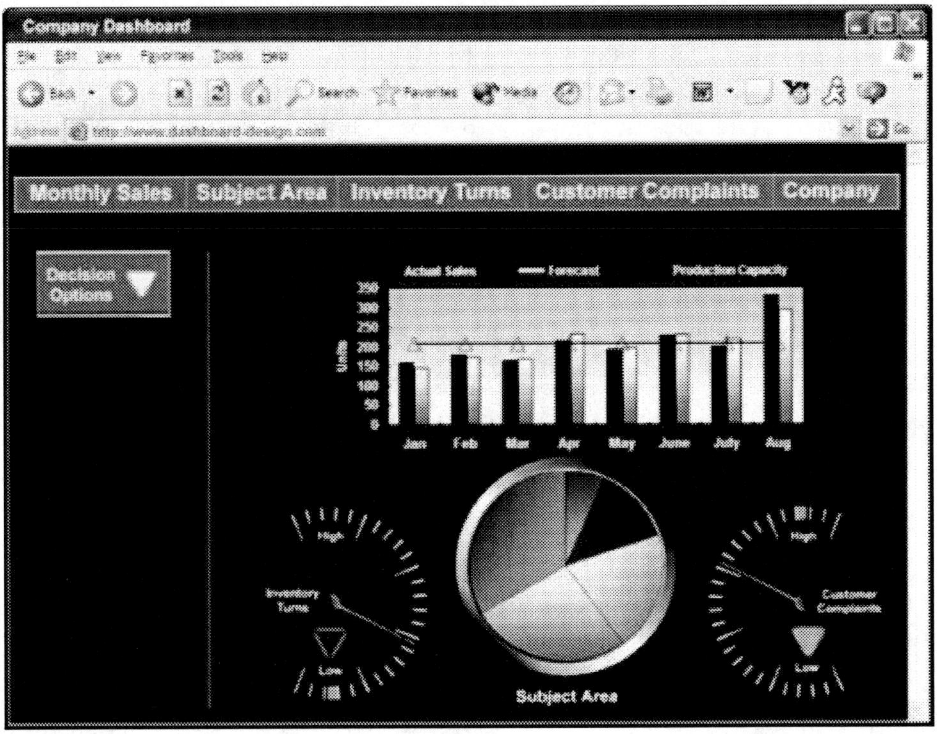

Figure 1.1: An example dashboard.

Dashboards don't often drive direct decisions; rather, they let individual users get a feel for general performance, such as sales, inventory turn, customer complaints, and so forth. Dashboards let you know if everything is all right or if further investigation is warranted. Dashboards may also include trends—such as day-by-day sales figures compared with goals or plans. Again, the dashboard isn't going to let you know *why* sales are where they are—but if they're significantly off-plan, you'll be able to tell at a glance and initiate further investigation.

Dashboards are useful because they can help new and less-experienced users quickly start taking advantage of the data in a data warehouse—or even the data in a transactional processing system, such as an order-entry application. The learning curve for a dashboard is usually pretty short and shallow, so more users are likely to use them more effectively.

Another useful visualization is a *scorecard.* This custom UI links internal and external data to the organization's goals. Basically, it tells you how far along you are in terms of achieving your goals. Figure 1.2 offers an example, showing how far various departments are in achieving a specific company goal.

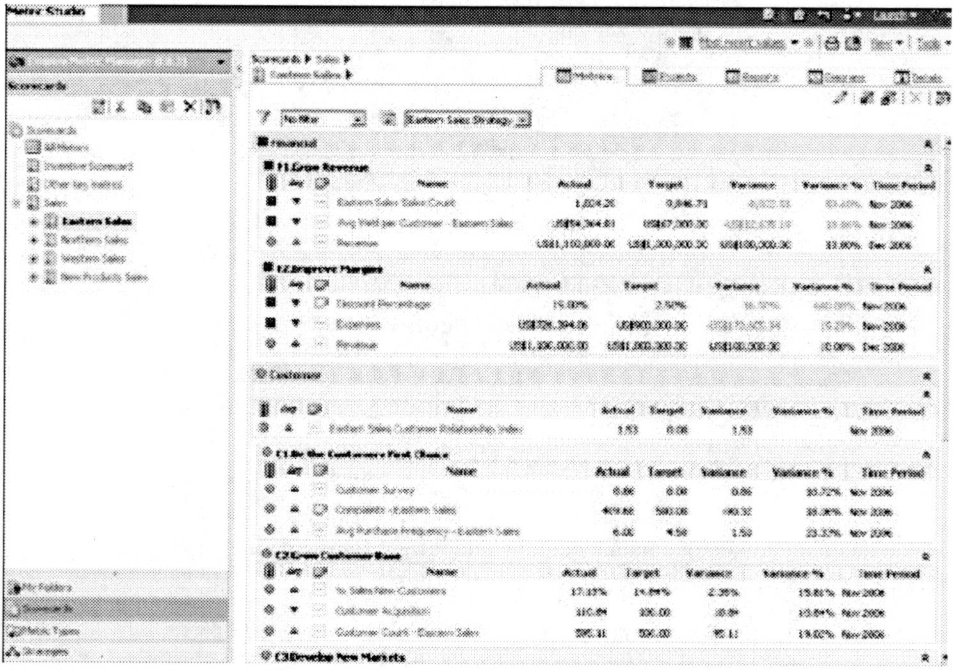

Figure 1.2: An example scorecard.

Scorecards are useful precisely because they show you where you are in relation to your goals. In this example, the Sales department is doing well; the Human Resources department may need some investigation and assistance. Scorecards help you focus your efforts, spot problem areas, and manage *to your goals* on a daily basis.

Dashboards and scorecards serve a similar purpose, and some BI solutions present them in such a way that there's no practical difference. If you need to make a distinction between them, a dashboard usually just shows you where you are in absolute terms; a scorecard shows you where you are relative to your goals. Both are useful, and both may be presented in a "dashboard-style" UI.

Empty Beds?!?!

Consider this example of how dashboards can help management focus on the right areas for their attention. Nice

Hotels, Inc. traditionally managed itself like most hotels do—their primary metric was room occupancy. 100% occupancy was always the goal, although most of their properties rarely achieved that on a consistent basis.

After implementing a BI system, managers were given a new dashboard that showed them total occupancy as well as occupancy on a *per-bed* basis. The dashboard also showed the number of guests turned away in a "no vacancy" situation. They found that their front desk agents were assigning single guests to rooms with double beds; later in the evening, they would turn away families simply because the only rooms left at that point contained single beds.

Managers started managing to *per-bed* occupancy instead of per-room occupancy. The scheduled room cleanings so that an equal number of doubles and singles would be cleaned at roughly the same time, ensuring that the front desk would have the best chance of having the "right-size" room for whatever guest was checking in. They trained the front desk to not use double rooms for single guests.

Over time, some properties were still seeing higher-than-desired quantities of empty beds each night, so they began planning to convert some double rooms to singles during refurbishments. Single rooms require less time to clean, are less expensive because they contain less furniture and soft goods such as sheets, and those properties clearly had a surplus of double rooms.

By being able to see per-bed occupancy at a glance, managers began to manage more aggressively and intelligently, customizing each property to its historical clientele and realizing additional revenue and savings over time.

Predictive Analytics

Predictive analytics is exactly what it sounds like: mining data, looking for patterns, and making predictions about future events based on historical facts. Credit scoring is one of the most well-known forms of predictive analysis, and credit reporting companies make use of some of the world's largest data warehouses. The reporting companies' proprietary scoring algorithms consider factors such as credit history, payment history, loan applications, and other customer data to assign numeric scores that are a prediction of the customer's ability to properly manage and service their debt.

Cross-sell—such as the "you might also like" product suggestions from e-tailers like Amazon.com—is another example. By analyzing the total purchases made by all of its customers, a company can predict, with some degree of accuracy, what products you are likely to purchase based upon those you have purchased or those you are considering. By offering you those items rather than making you go hunt for them, the company can take advantage of impulse buying, helping to increase overall revenue.

Internally, many companies use predictive analytics without even thinking about it. For example, many retailers know that their holiday sales will be some multiple of their pre-holiday sales; they also know that a certain number of additional employees will be needed to handle the additional holiday sales volume. By looking at historical data and plotting a trend into the future, those companies can make an informed guess about each year's holiday hiring needs. Most companies face numerous ad-hoc decisions every day; predictive analytics can't *make* those decisions, but it can help *inform* them—with more accuracy than a "gut" decision.

Business Performance Management

Business Performance Management (BPM), or Corporate Performance Management (CPM), rolls up reporting, predictive analytics, and other BI practices together with planning, budgeting, and forecasting. It's designed to provide a framework that organizes information, delivers new insight, takes action, and optimizes performance.

In "The Next Generation of Business Intelligence: Operational BI," author Colin White described the link between BI and business performance management:

> *The biggest growth area in operational BI analysis is in the area of business performance management (BPM). Operational BPM applications not only analyze the performance... but also compare the measured performance against business goals and alert business users when actual performance is out of line with business goals.*

As with scorecards, *goal* is the operative keyword for BPM. BI is great; comparing the information from BI to your goals allows you to *manage.*

Although the fancy name is new, BPM as a concept isn't all that new. Most sales managers, for example, are in the habit of looking at the prior days' sales when they sit down for work in the morning. They might drill into each salesperson's individual sales as well. They compare all that data with their sales goals, and typically know how much they still have left to sell in the month before they meet their goal. BPM and BI simply take that to a higher level, automating the collection of the data, automatically comparing it with goals, and not only displaying the result in a dashboard but also alerting management when performance is out of line with goals.

BPM might even drive automated decisions. For example, if sales are markedly higher than planned, a BPM system might accelerate the velocity of products orders placed with vendors to ensure sufficient stock is on hand to meet the growing demand trend.

BPM is a continual loop, meaning a BPM system updates itself. As the company responds to current events, the BPM system continuously collects its data and changes its analyses, helping the company immediately see the near-term and predicted long-term effects of its efforts.

A Brief Technical Overview of BI

Now, let's spend some time looking at the technology that lives under the hood of a BI solution. Many of these technologies are really just clever extensions of proven, decades-old technologies used in new ways to help achieve greater results.

Data Marts and Data Warehouses

A data warehouse—or a data mart subset—is really just a normal database. Typically, they live in the same relational database management systems—such as Oracle, SQL Server, DB2, or whatever—that "normal" databases live in. However, data warehouses and data marts are structured quite a bit differently.

OLTP vs. OLAP

An Online Transaction Processing (OLTP) or transactional database—has several key features:

- It contains detailed information. For example, it contains not only summary information such as the total amount of an order but also detailed information such as how much each item costs.

- It is designed to process transactions—meaning it's typically dealing with one piece of data at a time: one order, one product, one customer. It may be used to generate basic reports from this data, but it's structure is optimized to support rapid access to small chunks of data, therefore containing many tables.
- The schema is rigid and unchanging.
- It contains up-to-date information and is updated in realtime. It may purge data once it is no longer needed or relevant, therefore containing only current data.
- The quality of input data is often very high, meaning applications and other elements ensure that correct data goes into the data warehouse.

A data warehouse is an Online Analytical Processing (OLAP) database, and often includes these features:

- Some data may be summarized, meaning detail is not available. You may be able to tell the total amount of a given order but not the cost of each product contained in the order.
- Its purpose is to drive analysis and decisions. Access is optimized for examining large quantities of data in order to see trends.
- The information is historical and is only as up to date as the loading schedule. The emphasis is on past data and trends, more so than immediate, realtime data. Historic data from months or years ago will often be stored within the database.
- Data is often "cleansed," meaning errant or "edge" data may be removed to make trends clearer. Data warehouses tend to focus on sets of data rather than on individual elements.

Companies with a data warehouse will always have one or more "normal" databases that feed the data warehouse.

Normalized vs. Dimensional Data

OLTP systems rely on *normalized* data, meaning they seek to reduce data duplication and rely on dependencies between sets of data. Customers place orders, for example, and so an OLTP database might have one table for customer data and another for order data and would use relationships to connect customers to their orders. OLTP databases work well because in many situations only small pieces of data are being manipulated—entering a new customer or retrieving one order, for example. OLTP databases are tuned to support this behavior.

However, when it comes to BI, you don't always need the fine details—like which customers ordered what products. Instead, you're looking at bigger trends—like how many of a particular product were sold in the preceding quarter. OLTP databases can provide that information, but they're a bit slower at it because their database structure isn't fine-tuned to summarize and aggregate data in that fashion. In fact, *asking* an OLTP database to provide that information can impact performance of normal database operations—meaning your quest for BI will actually slow down business.

A data warehouse uses a different database structure entirely. Figure 1.3 shows a simple example of a *star schema,* one of the simplest data warehousing structures.

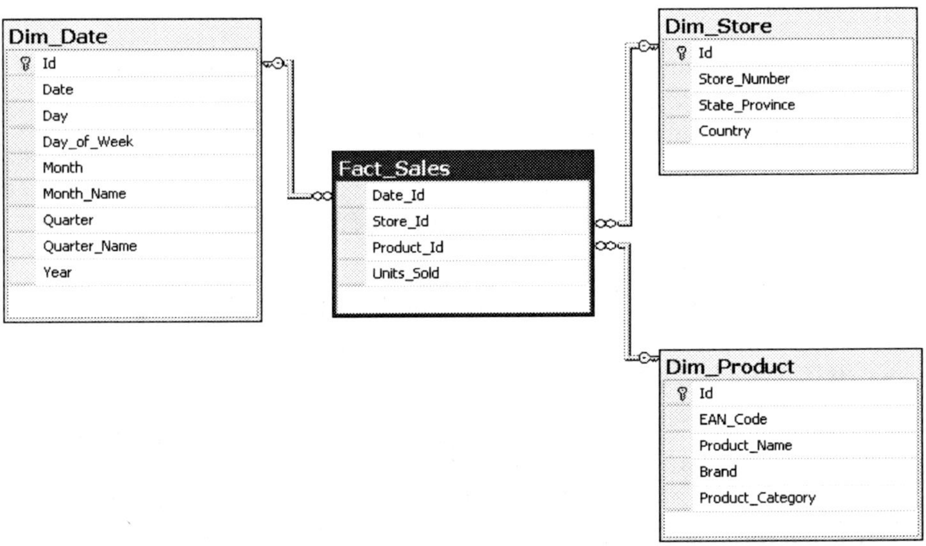

Figure 1.3: An example star schema.

Here, all the possible dates are pulled into one table—called a *dimension* table. All possible products are in a second dimension table, and all the company's stores are in a third dimension. A central *fact table* links the three dimensions so that you can discover which products were sold in which store on what date. This structure is not optimized to reduce data redundancy—in fact, there will likely be a lot of duplicated data. That's okay; data warehouses trade size for speed, meaning they often contain a great deal of redundant data, but that redundancy helps them produce results like reports much more rapidly. Using a schema like this, for example, makes it much easier to see the days on which televisions sold more quickly or whether there are any stores that sell laptop computers especially well.

Data Modeling

The trick to an effective data warehouse is in the schema—how you *model* the data. You need to know in advance, to a degree, what questions you want answered by the data warehouse. For example, the example in Figure 1.3 won't help you figure out whether more customers use Visa or American Express to purchase refrigerators because you didn't include payment information in the schema. If that's the type of thing you want to know, it needs to be included in the data model.

Data modeling is where large companies spend a lot of time when they're implementing a data warehouse. Fortunately, midsize companies often share a lot of common data sources and business needs. Just as an off-the-shelf bookkeeping package tends to work well with just about any kind of midsize business, an off-the-shelf BI solution can also work well—without all the time-consuming up-front data modeling.

Data Transformation and Loading

With your data warehouse schema ready, the next task is to get data into it. One reason a data warehouse works well is that it *copies* data from production systems into the warehouse; that means you can pull reports from the warehouse all day long without impacting your production systems. It also means that your warehouse's data will always be slightly out of date—exactly how much depends on how often you copy new data into it. However, most BI works from long-term trends, so not having up-to-the-moment data isn't usually an issue. There *is* such a thing as realtime BI, but it is beyond the scope of this book.

The actual process of copying data is usually referred to as an *extract, transform, and load* process (ETL). The extract step connects to the source database and pulls the required data from it. A good BI solution will have the capability to connect to most common databases; long-standing connectivity standards, adopted by most database vendors, makes this easier. The transform step rearranges the data into the schema used by the data warehouse. This may involve summarizing certain pieces of data, if the data warehouse won't contain line-item details, and it usually involves spreading the data out into the different table structure used by the data warehouse. Finally, the load step actually places the transformed data into the data warehouse. This may be a separate database running on a relational database management server or a proprietary database contained within the BI solution itself.

ETL isn't an all-or-nothing process. You may load data from certain sources once an hour, or every night; other sources—such as accounting systems—may become involved on a monthly or quarterly basis.

In-Memory Analytics

A relatively new development in the field of BI is *in-memory analytics.* Rather than copying all your data to a different location—like a data warehouse—and performing analysis there, data is simply read into a server's memory and analyzed right then and there. Actually, the *idea* of in-memory analytics isn't new, but it's only recently—with the availability of more powerful computing resources at lower prices—that it's become practical. I'll discuss it in more detail in the next chapter.

Reporting, Analysis, and Visualization

The last step, of course, is to *use* the data in the warehouse. This is where BI systems really come into play: Simply having a lot of data sitting in a database isn't terribly useful; the BI system contains the smarts to turn that *data* into useful reports, dashboards, scorecards, and other forms of *information.*

Good BI systems will allow users to work with the tools that they're already comfortable with. I've described how dashboards and scorecards provide intuitive summaries of data; some users may prefer straightforward reports, while others might prefer to work with pivot tables in a spreadsheet application such as Microsoft Excel. Figure 1.4 shows an example of these, and how a single BI solution can provide all these forms of output.

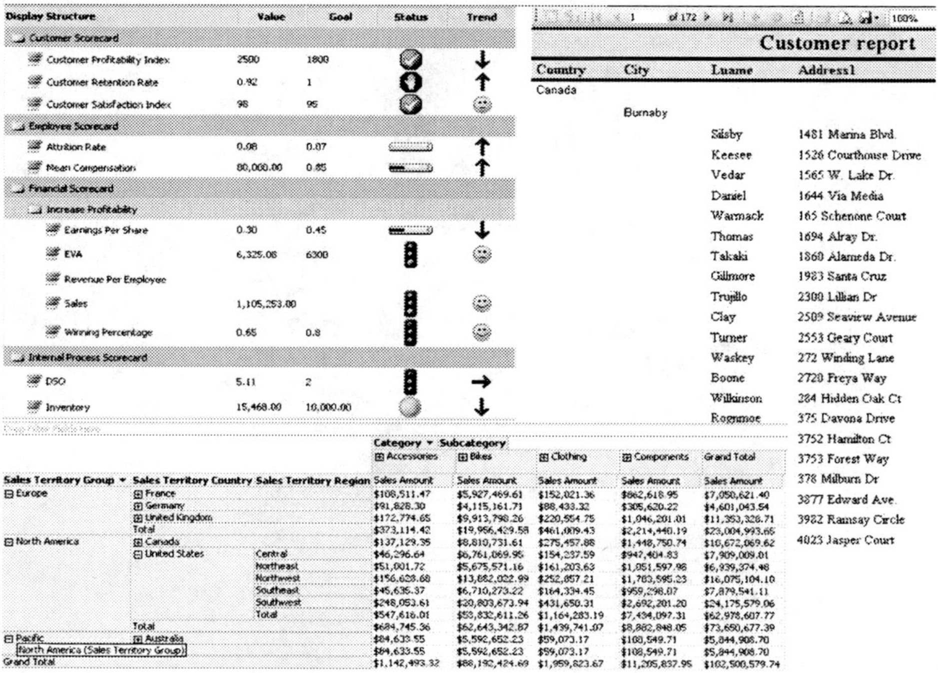

Figure 1.4: Common examples of data warehouse output.

Pivot tables—the "PivotTable" feature in Microsoft Excel, for example—can be useful analysis tools for users already comfortable working with spreadsheets. We'll discuss them more in later chapters, but for now, keep in mind that pivot tables allow users to construct custom output, summarize key data on demand, and rearrange data to see different relationships.

BI systems can also provide more powerful visualizations for data. For example, Figure 1.5 shows a relationship chart, which helps visualize patterns in data. In this example, the yellow nodes represent individuals in a community who smoke cigarettes. Researchers used this chart to study the effects of the smokers' relationships on their smoking, and found that after almost 30 years, most stopped smoking. The ones who continued smoking had few close relationships. This suggested a pattern of quitting—if your friends quit, you will too. It's a type of relationship that only a data warehouse, and this specific kind of visualization, can unveil.

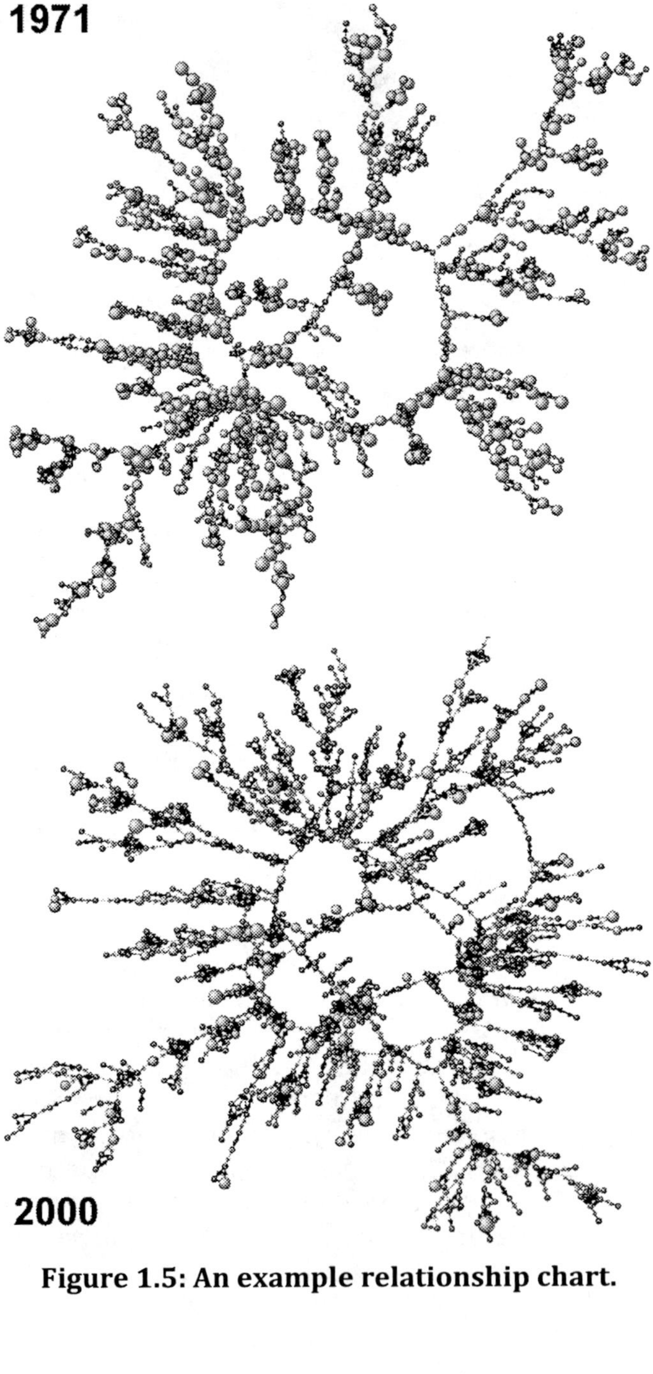

Figure 1.5: An example relationship chart.

Reporting, analysis, and visualization can all seem like the same thing—and they certainly all serve similar goals. So what are the real differences?

- **Reports** can address a variety of business needs, can be personalized and targeted, and can be easily distributed via many means. They can be adapted to run from different data sources as well, making them very flexible.
- **Dashboards** translate complex information into high-impact, at-a-glance displays. They're more intuitive than reports, and help keep decision makers aligned with organizational goals. Dashboards help you spot problems more quickly, making them good daily-use tools.
- **Scorecards** also deliver at-a-glance information but show how well you're doing compared with specific targets. They're good for communicating strategy and progress, and can help increase accountability.
- **Analysis** lets you explore information from across the business, and helps you dig deeper to discover the "why" behind what you're seeing on a report, dashboard, or scorecard. You can move through summary information to deeper-level information, and model "what-if" scenarios to help find the right action for improvement.

BI: Analysis + Planning

All the business "intelligence" in the world is useless if you don't *use* it to change something about your business—or at least to validate that what you're currently doing is the best path. Good BI solutions facilitate this goal by not only providing intelligence but also by helping you analyze that intelligence, make decisions, and actually implement those decisions.

Analyzing and Reporting on Information

Analyzing information can happen in many ways. You might use reports—or, more specifically, you might use an interactive *drill-down* or *drill-up* report. These reports let you click an item of data to see the detail behind that item, making it easier to dig to the source of something that is problematic. Or you might look at information based on ad-hoc queries, which can be presented in a variety of formats. You might also use dashboards or other visualizations, like the breakdown shown in Figure 1.6.

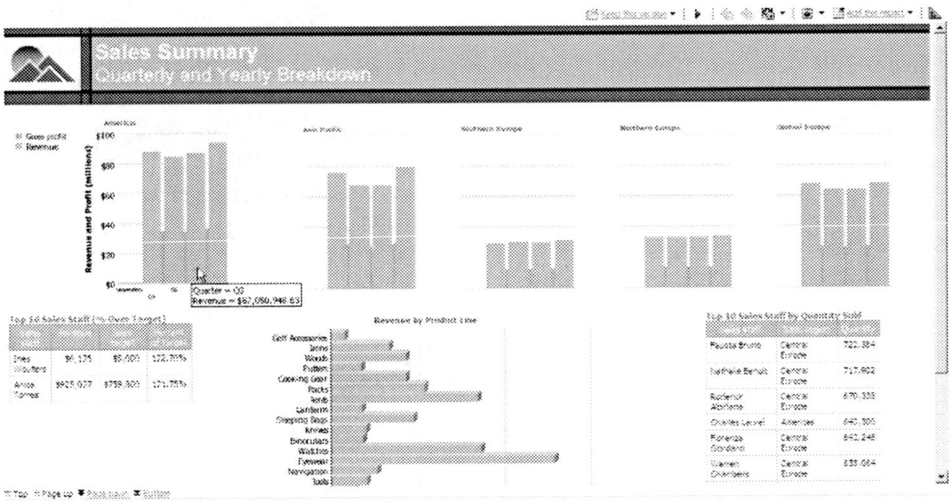

Figure 1.6: An example sales summary report.

If you're comfortable in Excel, you might create a PivotTable, or use one of the many BI add-ins for Excel. These can help you analyze "what if" scenarios, like seeing the effect of a change in product sales for a given month. All this analysis will help you draw conclusions about why things are the way they are within your company; you can then make informed decisions about which things need to change in order to meet your goals.

Making Decisions

Making decisions can be the tough part. The trick is to make sure you're looking at *all* the data. BI solutions can help by correlating data from many sources, making it easier to see previously-siloed information in a single, cohesive view. See product sales next to manufacturing costs alongside payroll and facilities overhead. Drill-down reports help you see the nitty-gritty details easily and then "drill up" to summary views that show you the "big picture" results of proposed changes.

"What if" is where BI really comes in handy. As you look at projections and trends, BI solutions make it easy to plug in different numbers: What if we increased payroll and hired new shipping clerks? What if we use an additional vendor to source critical components? What if unit sales go down 10%? What if we offer bigger discounts for bulk orders? By punching in proposals, you can see—almost instantly—how the numbers fall out, to determine whether your proposals make a positive or negative impact.

Implementing

Implementing is where the rubber meets the road: It's where you take your best proposals and put them into action. Doing so may require updating corporate budgets, financial outlooks, and other planning and support systems; a good BI solution will provide *write-back* capabilities so that your "what if" proposals can be "accepted" and linked back to business optimization and planning tools. In other words, just as a BI solution can *read* data from planning systems, it can also *write* data back into those systems to adjust assumptions and plans. It's a great way to "close the loop" in BI, making your "what if" proposals a reality, sooner.

One Size Does Not Fit All

BI—like most business technologies—is not a one-size-fits-all affair. I actually get a little surprised when I see midsize businesses attempting to implement—or even considering—the same BI solutions used by a giant company like Ford Motor Company, Aetna, or Home Depot. Most midsize companies would never even look at the ERP, CRM, and other applications suitable for a giant company; most solution vendors have specific products geared toward midsize businesses. Why should BI be any different?

BI in Large Enterprises

No two large enterprises are exactly alike—indeed, few of them are even vaguely similar, even when they're in the same industry. They may have all started as small companies, but they grew up in very different ways. They handle payroll differently, structure their accounting differently, have different manufacturing models and philosophies, and so on. Large enterprises are often organized into business units and divisions which themselves operate almost as independent entities. In some large enterprises—GE is a good example—there are divisions that have absolutely nothing to do with other divisions, such as GE's medical equipment division and their television broadcasting division (although GE's medical equipment does get suspiciously good placement in their networks' television shows).

All of this means that BI in a large enterprise is *complicated.* You're dealing with thousands of data sources in the ETL process and may have hundreds of different audiences that need different reports, dashboards, and scorecards. Every BI deployment in one of these enterprises is a custom affair, from designing the data warehouse schema to writing the ETL routines to developing the final output. Many BI vendors who work at this level don't sell "products;" they sell *toolkits,* along with deployment and implementation services—experts who use those tools to build a custom BI solution for each new customer.

Sound expensive? Sure—but don't forget that big enterprises have a lot of money on the line. If they spend a couple of million dollars implementing a BI solution, they may be looking to save *tens* of millions from that solution. But it *doesn't scale down.* Follow that same implementation process, using the same vendors and toolkits, and you'll spend a lot of money—and midsize companies aren't looking to *save* as much, so the BI investment doesn't seem worth it.

BI in Midsize Companies

Midsize companies are a totally different animal. They're more likely, for example, to use off-the-shelf or lightly-customized solutions for key tasks such as bookkeeping, payroll, CRM, and so forth. Midsize companies tend to focus on similar broad business questions: Where are sales? How's inventory? What does payroll and other overhead look like? Midsize companies tend to have a lot of broad similarities with one another, in other words.

Thus, BI vendors can offer prepackaged BI solutions designed for midsize businesses. These solutions may require little or no outside expert services to implement, and they're designed in much the same way that midsize bookkeeping software is designed—to meet a common set of needs, with enough room for customization to ensure a good fit on most businesses. The BI vendor is able to spread the cost of designing the data warehouse, output, and ETL routines across all of its customers, resulting in a lower-cost, off-the-shelf product. It wouldn't be at all suitable for a Fortune 500 company—but it can work great for midsize businesses.

Prepackaged BI solutions often offer the most compelling features from the "full-sized" BI toolkits, such as Web-based dashboards, support for pivot tables in spreadsheets, scorecards, and so on. We'll explore some of these features and help build a BI solution shopping list in an upcoming chapter.

Coming Up Next...

So exactly what is involved in making BI happen inside your organization? How do you introduce a BI solution, and what sort of outputs will you need? How will you analyze that output, and what sorts of decisions can you expect to make? In the next chapter, I'll attempt to answer all of these questions by explaining the basic process by which BI is introduced into an environment.

We'll dig a little deeper, technically, than in this chapter, but I'll also focus on laying out the steps you'll be taking. I'll look at why BI is traditionally complicated and expensive, and outline some ways in which midsize companies can leverage certain economies of scale and similarity to save both time and money.

Chapter 2: How Business Intelligence Happens

How, exactly, do you "get" a business intelligence (BI) system into your organization? BI hasn't traditionally been something you just install onto a server or client computer, so it's important to understand the process, tools, and techniques that are involved in implementing and creating it. Some of these may, in fact, be major reasons why your midsize company has avoided BI in the past—making it even more important to understand not only the technologies, tools, and techniques but also how they've evolved in recent years to meet the needs of companies other than giant enterprises.

In this chapter, I'll explain the basic processes by which BI is introduced into an environment. I'll also look at some of the reasons BI is traditionally a time-consuming and expensive proposition for most companies, and set up some of the ways in which you can implement BI more easily and for less money. Think of this chapter as the "BI life cycle"—a look into what BI actually looks like inside an organization like yours.

Assembling Your Business Data

BI is, first and foremost, all about data. The ultimate purpose of BI is to bring together data from many different sources so that you can start making connections, finding patterns, and spotting trends that wouldn't be apparent otherwise. That means the first step in implementing BI is deciding what business data you'll need to assemble, and figuring out how to get it all into one place. The actual technologies for doing so are less important right now; we're concerned with the different *kinds* of data that drive your business and the ways in which that data is currently *stored* and *used.* That information will help us decide which techniques you should use to pull that data into a BI platform.

> **Note**
> I'm deliberately avoiding the term "data warehouse" for a couple of pages; although it's likely that your BI effort will involve some kind of data warehouse or data mart, that's not important at the moment. Right now, we're taking a step back from the deeper levels of BI technology and thinking about what your data currently looks like, and what you need your data to look like in order to use it within a BI implementation.

Right now, the majority of the data in your organization probably lives inside *transactional databases.* That is, your data lives in the systems that handle your business' day-to-day operations—customer ordering systems, invoicing systems, financial systems, and so on. That information might physically live in something like Microsoft SQL Server, or Oracle, or IBM DB2, or some other relational database management system. You may well have data living in smaller databases such as Microsoft Access or even Microsoft Excel spreadsheets. Some of your data may be in proprietary databases, like a QuickBooks file, or in a database used by midsize business software like Microsoft Dynamics ERP. These are called source systems or systems of record.

BI doesn't change that. Your day-to-day data will continue to live wherever it lives now, driving business applications, line-of-business software, and the other systems that make your business work. What BI wants to do is *look* at that data, possibly *copy* that data to different locations and in different forms, and to *analyze* that data.

Direct Data Mapping

Direct data mapping allows a BI system to connect to your business data in its current location. That is, rather than copying your data somewhere else (like into a data warehouse), the BI system simply connects to the data whenever it needs to use it. The advantage of doing so is that your minute-to-minute changes can be immediately viewed and used within the BI system; a disadvantage is that—depending upon how your BI system is built and used—you can place a significant performance burden on your live production systems, something that you'll typically want to avoid.

The *mapping* part of this refers to the fact that BI systems—as discussed in the previous chapter—typically organize data differently than a transactional system does. BI systems use data structures that are optimized for fast querying of data, whereas transactional systems often try to strike a balance between querying data and adding new data. In addition, transactional systems rarely execute the kind of complex, interrelated queries that a BI system uses.

Interestingly, data mapping almost always comes into play with a BI system—even if you don't want to allow your BI system to query data directly on-demand from transactional systems. Sometimes, you might prefer to *copy* relevant data from your transactional systems into a dedicated BI data store; when you do so, you'll still need a data map that helps the BI system properly query and re-organize the data for use within the BI solution.

> **Simplified**
> *Data mapping* is the process of designating the data that will be utilized in the BI system, whether that data is accessed directly where it sits or copied into a different data store for BI purposes. Data mapping designates not only that data that will be used but also—to a degree—how the data will be re-structured into the form used by the BI system.

Data Loading

Data loading is used when you *don't* want a BI system to directly access production data while users are trying to enter orders, find inventory, and balance accounts. Utilizing a data map that is either manually created or, in some cases, generated automatically by software tools, data-loading utilities copy data out of your transactional systems, *transform* or rearrange the data into the structure required by the BI system, and add the data to the BI system.

In the BI world, data loading is often referred to as *Extract, Transform, and Load,* or ETL, which nicely describes exactly what happens: Data is extracted from one database, transformed into a different structure, and loaded into the BI system. Some BI systems may be pre-programmed with data maps for commonly-used business software, including Enterprise Resource Planning (ERP) solutions or other financial software; most BI systems also allow custom data loading from whatever databases you happen to have in your environment.

The *transform* aspect of ETL is what helps make BI so useful. Because different databases all use different structures and conventions, connecting data from different places can be very difficult. Transforming the data helps standardize it. For example, a transform process might:

- Translate abbreviations, like the names of states, into their full names
- Select only certain pieces of data that will be relevant to the BI system
- Cleanse data, such as running customer addresses through an address verification system or formatting all product ID numbers to a standardized format
- Aggregate data, such as calculating the total for custom orders, even though your transactional system only stores line-item amounts
- Filter data, eliminating redundancies or unwanted data
- Joining data from multiple sources, such as looking up product names rather than storing less human-friendly product ID numbers
- Validating data, ensuring that the data used for BI purposes is correct, valid, or up to date
- Correlating data from different systems—matching the data from your order entry system to the data in your inventory system to the data in your shipping system to the data in your customer relationship management (CRM) system

Data loading is often accomplished during off hours, when any burden it places on your production systems will be less noticeable to your users. This means that your BI system won't have up-to-the-minute data—but in most situations, that's okay; BI systems are all about working with longer-term trends and patterns, and less about working with up-to-the-second data. However, if your situation requires that your BI system work with some kinds of real-time data, direct data mapping can be used in conjunction with data loading to provide a hybrid model that gets you what you need, when you need it.

Data Marts and Data Warehouses

As discussed in the previous chapter, data warehouses are typically where a BI system stores its data—after loading that data from your production, transactional systems. To quickly review, a *data mart* commonly holds data related to a *portion* of your business, such as departmental data. A data mart can't help make broad decisions that affect the entire company, but it can be used to help make decisions within the portion of the company that it serves. A *data warehouse* contains standardized, consistent, cleansed data from the entire company (or from most of it, at least), and is used to analyze patterns and trends from across the company.

For example, a data mart might help your CFO understand the relationship between salary costs and product sales, and make "what if" decisions to help predict future salary expenses in various sales scenarios. A larger data warehouse, in contrast, might help tie in information on your marketing and advertising efforts, helping the CFO predict changes in salary costs based on different advertising efforts or marketing campaigns—as those things presumably affect product sales.

Some data warehouses are built from a collection of data marts (called a *bottom-up* approach); other experts prefer to build the larger data warehouse first, then break it down into departmental data marts as needed (the *top-down* approach). There are strong opinions held by different experts for both techniques, and which approach to take is often one of the first tough decisions that large companies face when contemplating a BI implementation.

Top-Down, Bottom-Up... Help!

Pick up a good book on data warehouse design and you may quickly come to the decision that BI isn't right for your midsize company. Design philosophies are just the beginning of the difficult decisions it seems you'll have to somehow make.

But that situation is actually true for most business decisions. Will your company operate on cash basis or accrual? How will you structure your financial ledgers and accounts? How will you build and monitor a sales pipeline?

Giant companies are often forced to confront these decisions, spend time educating themselves, and make a decision. The technology solutions sold to large companies must often be customized to fit the company's decisions, which can be expensive and time consuming.

Midsize businesses, however, often take a "commonly accepted" approach. They purchase an off-the-shelf accounting, ERP, or CRM system, then use it however it was designed to be used. In a way, they let the technology make the decision for them, because, in most cases, the technology was written to model the most common, "best practices," and "generally accepted" way of doing things. In this way, midsize businesses save themselves time and money.

BI should be no different. Although giant companies will still need to spend a great deal of time on design decisions and other up-front questions, midsize businesses can often select a BI solution that was *designed for midsize companies,* letting the solution's designers select the most common, generally-accepted way of doing things, then simply working within that framework rather than having to make every single design decision themselves.

In-Memory Analytics

One of the hot new trends for BI in the past few years has been *in-memory analytics.* The idea here is that rather than loading all your data into a data warehouse (which is really just another kind of database that stores data on disk), you load your data into a computer's memory and work with it there—interacting with disk-based storage very little if at all. In-memory analytics isn't actually a new concept but the recent (and massive) drop in price of computer processors and memory has finally started to make in-memory analytics more practical.

In-memory analytics can help reduce or eliminate the need to transform and load data into a data warehouse—something that is often time consuming—before analyzing the data. Instead, data can be queried from production sources, "transformed" in-memory, and analyzed directly. This is especially useful for obtaining more up-to-the-minute analyses and for analyzing data in a new way that isn't provided for in a traditional data mart or data warehouse.

The in-memory technique is important for a few reasons:

- Building a data warehouse requires technical expertise, which is also required to help tune the performance of the data warehouse. Because in-memory analysis is ad-hoc, less technical expertise is needed, allowing business analysts and line-of-business managers to build their own reports and dashboards.
- In-memory analysis is faster than analysis that requires disk access simply because disks are much slower than computer memory.

- In-memory analytics can allow more flexible analysis of the data. Data can be rapidly re-structured in memory to answer different questions without having a team re-design the data store and re-build the database tables to support it.

In-memory analysis isn't a perfect solution, though, because computer memory still isn't unlimited or free. Some estimates suggest that computers being used for in-memory analysis need to have multiple fast processors and must be able to dedicate a large portion of memory to the analysis—enough memory to hold as much as 100% of the amount of data being analyzed, depending on the exact circumstances. In-memory compression and other data-handling techniques can reduce that memory requirement in some cases; some vendors cite a 1-to-10 data-volume ratio, meaning you'd only need enough memory to hold about 10% of the data you plan to analyze. Midsize businesses—which often deal with less sheer volume of data than huge enterprises—may find that in-memory analytics is an affordable and practical technique for their BI needs.

In-Memory Advantages

Business users *hate* waiting for answers. Tell a manager that his reports will only come once a week or that he'll have to wait a few hours to get responses to analysis queries, and that manager will probably give up on BI. In fact, many organizations still relegate BI system use to a few trained, patient business analysts. But the power and flexibility gained from giving front-line managers access to fast BI tools can't be ignored.

With in-memory analytics, query time might drop from 30 minutes to a few seconds—provided you can employ the hardware needed to make it happen. Once users become accustomed to such rapid performance, they start asking the BI system more questions, getting more answers, and running the business based on facts and trends rather than on gut instinct.

In-memory analysis can work without a data warehouse, meaning you won't need as much on-staff technical expertise to get up and running with BI. That makes in-memory analysis especially attractive to midsize companies whose IT staff might not have BI expertise or who simply might not have the bandwidth for a more complex BI project.

In-memory analysis has started becoming popular not because if its advantages—after all, who wouldn't want faster answers?—but because of its feasibility. Just a few years ago, 1GB of computer memory cost over $150; today it costs less than $35 or so. A powerful analytics server with 64GB of memory might have cost $64,000, and today costs only $13,000—much more affordable. The increased power—and lower cost—of 64-bit, multi-core computer processors has helped too, as those processors can physically connect to the large amounts of memory needed to make in-memory analytics possible.

Resource

If you'd like to read more about in-memory analysis, visit http://www.tagonline.org/articles.php?id=298. You'll find a great, clear article on the subject.

Creating an Information-Driven Business Culture

"Every time we raise the price of the product, we sell more copies." That's what one business executive told me, and that simple correlation drove many of the company's business decisions. Of course, *correlation* is not *causation;* the company was also adding significant new features to the product, exhibiting at more trade shows, and spending more on marketing and advertising. But that one simply factor—product price—is what drove that executive's gut decisions.

Managing by instinct is especially common in midsize companies, particularly because many executives and managers have been with the company for a long time and still feel they have a complete view into all of the company's data. But as tiny companies grow to be midsize ones, it becomes more and more difficult for a single human being to keep their fingers on every one of the company's pulses; managers stop managing based on information simply because they don't actually *have* all the available information. Eventually, they simply start managing on instinct—something that is unreliable, not repeatable, and generally doesn't serve midsize companies well in the long run.

After you get your data into a BI system, you have to start teaching yourself and your colleagues to *use* that information and to *trust* it. Create an *information*-driven culture— rather than an *instinct*-driven culture—in your midsize company. BI systems offer a number of efficient—and, frankly, sometimes cool—ways of doing so.

> **Skill ≠ Instinct**
>
> Not managing by instinct is *not* the same thing as not needing managerial skills. Good managers can still rely on their experience and their intimate knowledge of how the company works—that experience and knowledge can simply be more *informed.* Providing facts and information, filtered and tempered by experience and skill, is the way BI can make managers' decisions better.

Of course, I have to acknowledge that creating an information-driven culture does *not* mean that experience and hands-on management aren't useful. I've known—and in some cases worked for—too many managers that spent all day staring at spreadsheets and never tempered all that information with a hands-on reality check. They weren't, to put it kindly, all that effective at managing. Information should *contribute* to the management process, not dominate it.

Reporting

Good managers love reports. Reports offer information—data points that can help direct management decisions. A BI system simply produces broader, more insightful reports than those offered by a single transactional system. An order-entry system, for example, might tell a manager the average value of all the orders entered for a typical business day. A BI system, however, can also tell that manager the average cost to enter those orders because the BI system can draw from many other sources of data—including sources that offer salary amounts, utility bills, and much more. A BI system can also help that manager see trends in order volume based on advertising; a skilled manager backed by a good BI system will know exactly how many additional staff will be needed to handle the onslaught of customer orders that will accompany the newest upcoming television ad.

Reports from a BI system can be highly-flexible, and even dynamic, allowing managers to use a report to ask "what-if" questions, and make decisions based on the results. Figure 2.1 shows a typical BI report, displaying aggregate sales, sales by product category, and top-selling products. This kind of summary-level report helps managers focus their efforts, make product stocking decisions, and so on. A dynamic version of this report might be clickable, letting managers drill down into more detailed information as needed.

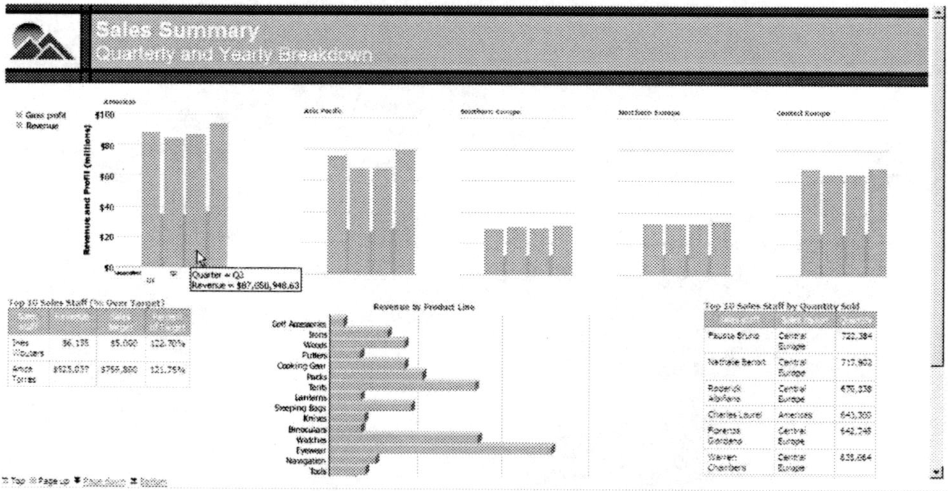

Figure 2.1: A common BI report.

Some managers are skilled in using Microsoft Excel, which offers built-in pivot table functionality and can use add-ins to provide even more BI analytical functionality. By connecting Excel to a data warehouse or an in-memory analytics server, as Figure 2.2 shows, managers can pivot data, build custom analyses, drill through data, and propose "what-if" scenarios to see the results. This helps managers see the potential outcome of decisions, helping drive the decision that most benefits the company. Figure 2.2 actually shows an Excel add-in, which is something that many BI solutions include or offer.

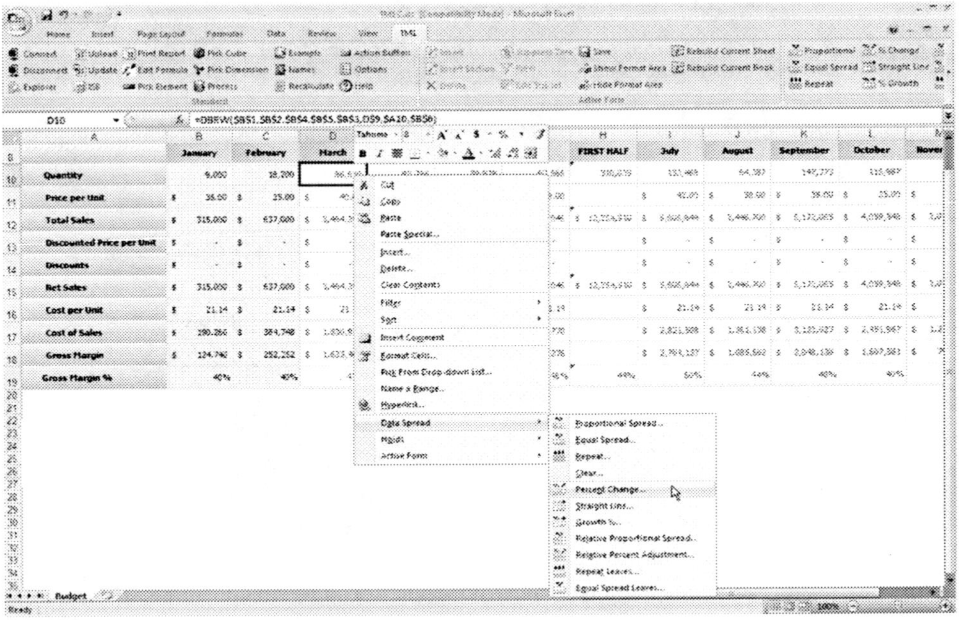

Figure 2.2: Connecting Excel to a data warehouse or in-memory analytics server.

The biggest challenge with a BI implementation is getting managers to *use* these powerful tools. Once exposed to them and shown how to utilize them, most experienced managers will never turn back.

Dashboards

The report in Figure 2.1 might be more accurately called a *dashboard,* which is a common feature of most BI systems. Dashboards are specialized kinds of reports that utilize graphical displays to summarize key information for at-a-glance consumption. Dashboards—like the dashboard in your car—help manager keep track of critical, high-level metrics. Figure 2.3 is another dashboard example.

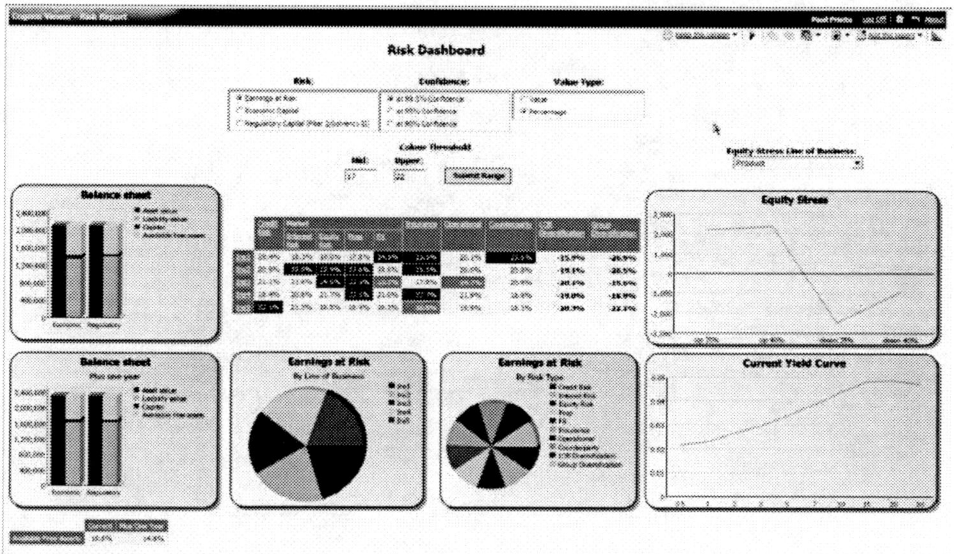

Figure 2.3: BI dashboard.

This dashboard might be ideal for executives or top-level managers, as it helps them see the company's current balance sheet, the projected balance sheet in one year, earnings at risk by line of business and by risk type, and other high-level metrics. Dashboards aren't designed to drive immediate decisions; like your car's "check engine" light, a dashboard indicator is intended to drive further investigation if something looks amiss.

Scorecards

Scorecards, like dashboards, are intended to give a high-level overview of key metrics. The big difference with a scorecard is that—like the ones you remember so fondly from school—a scorecard applies an absolute grade based on relative thresholds. In other words, for each metric, the scorecard tells you if the company is doing "good," "okay," or "bad." Figure 2.4 shows an example, where color-coded metrics help draw managers' attention to problematic areas—like sales growth, Web site visitors, and improvement suggestions, which are all red in this graphic.

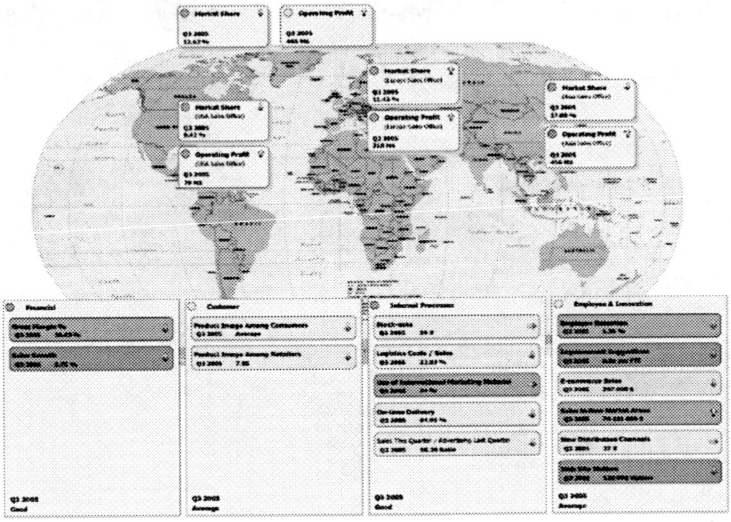

Figure 2.4: Example scorecard.

The idea of a scorecard is simple: If everything is an "A" or "B," then you're basically doing well. Items with a "C" or "D" deserve additional attention and managerial effort; items with an "F" are in need of immediate and significant attention. As with a dashboard, the idea is not to drive immediate decisions but rather to help focus effort and further investigation where they are needed most urgently.

Graphics = Engaging

Humans are visual creatures. Few managers relish the thought of staring at rows and columns of numbers, trying to pick out areas that they need to focus on. Informational graphics like those in dashboards and scorecards engage our visual senses and help us quickly sort out priorities.

A typical corporate profit and loss (P&L) statement is an example of how rows and columns of numbers aren't always good at driving managerial attention. Expenses are at 10%? Okay...is that good or bad? We spent $23,000 on shipping? Great...or not?

A dashboard, however, can show a chart comparing current performance with last year's; any bar in the chart that is significantly out of whack is an immediate call for attention. A scorecard can be designed to color-code anything that is more than 10% out of line, helping draw attention where it's needed most.

Although traditional textual reports will always have their place, one of the biggest and most immediate values a BI system can deliver is dashboards and scorecards that help managers focus their time where it is really needed.

Analyzing Business Information

Let's briefly pause and examine where we are in the BI life cycle. Figure 2.5 illustrates what we've already accomplished.

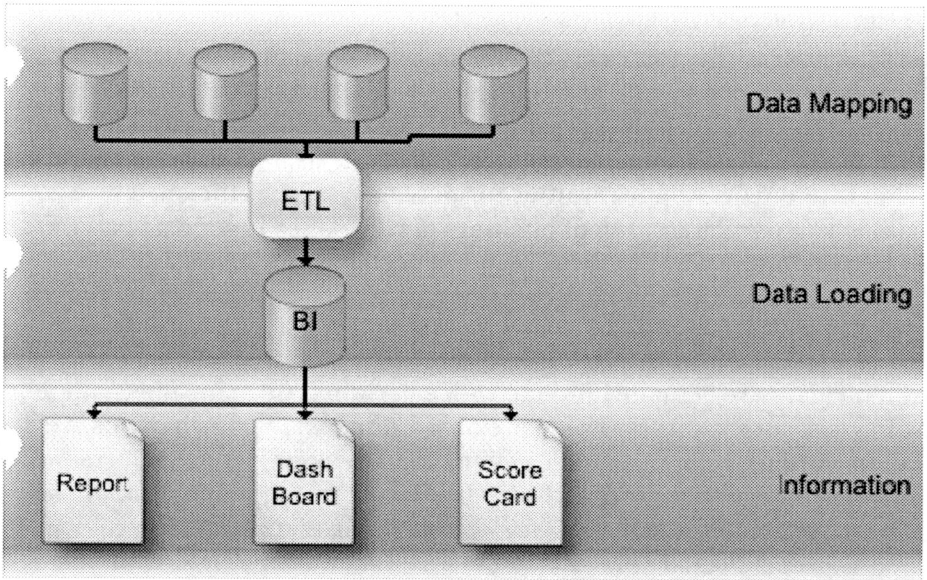

Figure 2.5: Our progress so far.

We've mapped data from live data sources throughout the organization. In some cases, our BI system will use that data directly, while in others, it will extract, transform, and load that data into the BI system—either into a data warehouse or perhaps into an in-memory analytical structure. From there, the BI system will produce the reports, dashboards, and scorecards that we'll be using to check on our organization's health and progress, ask "what if" questions, and so forth.

The next step is to get this information into the brains of managers, analysts, and others who will use that information to help drive business decisions. Should I draw an arrow leading from the information to a person? Probably not: One thing we have to acknowledge is that you can't *push* information to people; people have to voluntarily *pull* that information, actively examining it and using it to help inform their decisions. That's ultimately what makes a BI implementation succeed or fail: Is it used by the right individuals to inform the right decisions? Of course, you can help speed BI adoption by ensuring that your BI system provides accurate, timely information to the right people within your organization—and ensuring that it does so quickly.

Once that BI information is in the right hands, or rather heads, and they're prepared to use it, the act of analysis and decision-driving truly begins.

Analysis and Visualization

One of the most difficult aspects of a BI system is teaching its users how to analyze data. Simply staring at a report or dashboard isn't sufficient; BI users need to know how to "slice and dice" data to find the answers to specific business questions, see patterns, spot trends, and drive decisions. This helps to uncover opportunities for new markets or efficiencies, or to explain why a particular area of business performance is good or bad (and yes—you definitely want to focus on good performance as well as bad, because the causes of good performance can often be utilized elsewhere to improve the overall business).

Most managers are actually familiar with a certain type of business modeling and analysis: spreadsheets. These are a two-dimensional form of business modeling, often displaying metrics over time, breaking down performance by business categories, or something else. That's *two* dimensions: metrics and time, performance and category, and so forth. Figure 2.6 shows a common spreadsheet, displaying actual business metrics against planned metrics. Again, that's two dimensions: actual and planned.

Revenues Budget	Budget	Actual	Variances	%	Cum. Budget	Cum. Actual	Variances	%
Units 22785	450,000	510,334	60,334	13.4%	1,404,650	1,468,012	63,362	4.5%
Revenues 2	360,000	374,223	14,223	4.0%	1,059,900	1,074,010	14,110	1.3%
Units 3376	26,334	26,750	416	1.6%	74,115	77,025	2,910	3.9%
Units 44578	19,000	23,567	4,567	24.0%	54,500	70,246	15,746	28.9%
Total sales	855,334	934,874	79,540	9.3%	2,593,165	2,689,293	96,128	3.7%
Use right click & Unhide to show more rows.								
Cost of Sales	-255,033	-263,356	-8,322	3.3%	-776,664	-781,575	-4,911	0.6%
Gross Margin	600,301	671,518	71,218	11.9%	1,816,501	1,907,718	91,217	5.0%
Gross Margin %	70.2%	71.8%	-1.6%	-2.3%	70.0%	70.9%	-0.9%	-1.3%
Operating Expenses	-53,850	-60,423	6,573	-12.2%	-161,550	-179,690	18,140	-11.2%
Operating Profit/Loss	546,451	611,095	64,645	11.8%	1,654,951	1,728,028	73,077	4.4%
Overheads	-189,750	-60,841	128,909	-67.9%	-189,750	-183,636	-6,114	3.2%
Profit/Loss After Overheads	356,701	550,254	193,554	54.3%	1,465,201	1,544,392	79,191	5.4%
Operating Margin %	41.7%	58.9%		17.2%	56.5%	57.4%		0.0%
Expenses Budget								
Expense1	1,250	1,550	-300	-24.0%	3,750	4,450	-700	-18.7%
Expense2	2,000	2,015	-15	-0.8%	6,000	5,988	12	0.2%
Expense3	4,000	5,150	-1,150	-28.8%	12,000	15,595	-3,595	-30.0%
Expense4	4,500	3,445	1,055	23.4%	13,500	12,470	1,030	7.6%
Expense5	1,500	1,525	-25	-1.7%	4,500	4,415	85	1.9%
Expense6	30,000	34,034	-4,034	-13.4%	90,000	99,034	-9,034	-10.0%
Expense7	2,000	1,845	155	7.8%	6,000	5,729	271	4.5%

Figure 2.6: Analysis via spreadsheet.

The problem with two-dimensional analysis is that it's far too limited. Two dimensions aren't nearly sufficient to represent the complexities of even a very small business, let alone a typical midsize business. Although spreadsheets like this are useful for quick overviews, they don't provide an easy way to drill down into the data to uncover hidden details and root causes.

By contrast, BI is built around the concept of *multi-dimensional analysis.* It can be tricky to explain and illustrate, simply because computers are inherently two-dimensional when it comes to displaying information.

To display multi-dimensional information requires display tricks and conventions to mimic multiple dimensions; a spreadsheet *pivot table* is one such convention. Figure 2.7 shows an example pivot table.

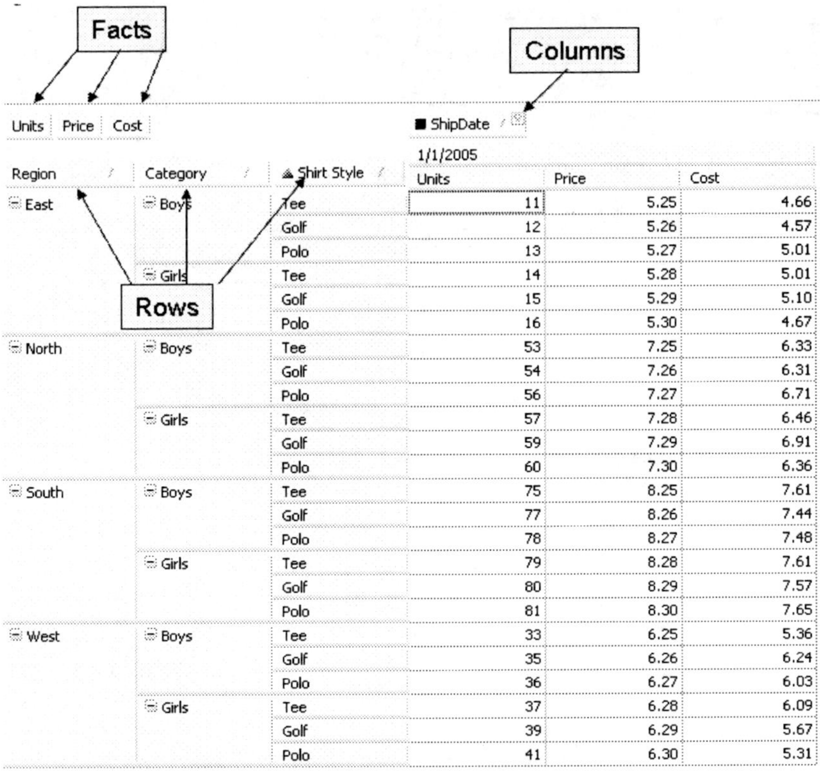

Region	Category	Shirt Style	Units	Price	Cost
East	Boys	Tee	11	5.25	4.66
		Golf	12	5.26	4.57
		Polo	13	5.27	5.01
	Girls	Tee	14	5.28	5.01
		Golf	15	5.29	5.10
		Polo	16	5.30	4.67
North	Boys	Tee	53	7.25	6.33
		Golf	54	7.26	6.31
		Polo	56	7.27	6.71
	Girls	Tee	57	7.28	6.46
		Golf	59	7.29	6.91
		Polo	60	7.30	6.36
South	Boys	Tee	75	8.25	7.61
		Golf	77	8.26	7.44
		Polo	78	8.27	7.48
	Girls	Tee	79	8.28	7.61
		Golf	80	8.29	7.57
		Polo	81	8.30	7.65
West	Boys	Tee	33	6.25	5.36
		Golf	35	6.26	6.24
		Polo	36	6.27	6.03
	Girls	Tee	37	6.28	6.09
		Golf	39	6.29	5.67
		Polo	41	6.30	5.31

Figure 2.7: An example pivot table.

> **Note**
> Microsoft's brand name for pivot table functionality within Microsoft Excel is "PivotTable;" the more generic "pivot table" describes this functionality in any product that offers it.

Pivot tables typically consist of multiple *facts*, which are added by the user. Here, the facts are Units, Price, and Cost—numeric dimensions. The columns and rows can be any kind of dimension: *Row* dimensions create a row for every unique value, while *Columns* create a column for every combination of column fields and facts. A simple pivot table like the one in Figure 2.7 can help spot critical business problems, such as poor unit sales in a given region or higher unit costs in a given category in a given region. It's those multiple dimensions—regions, categories, and styles—that help bring more information into play.

But pivot tables can reveal much more. In Figure 2.8, I've added a total row—shown in light yellow—to help reveal aggregate information.

Units / Price / Cost			ShipDate				
			1/1/2005			2/1/2005	
Region	Category	Shirt Style	Units	Price	Cost	Units	Price
⊟ East	⊟ Boys	Tee	11	5.25	4.66	16	
		Golf	12	5.26	4.57	17	
		Polo	13	5.27	5.01	19	
	Boys Totals		36	15.78	14.24	52	
	⊟ Girls	Tee	14	5.28	5.01	20	
		Golf	15	5.29	5.10	21	
		Polo	16	5.30	4.67	22	
	Girls Totals		45	15.87	14.77	63	
East Totals			81	31.65	29.01	115	
⊟ North	⊟ Boys	Tee	53	7.25	6.33	59	
		Golf	54	7.26	6.31	60	
		Polo	56	7.27	6.71	61	
	Boys Totals		163	21.78	19.35	180	
	⊟ Girls	Tee	57	7.28	6.46	62	
		Golf	59	7.29	6.91	64	
		Polo	60	7.30	6.36	66	
	Girls Totals		176	21.87	19.73	192	
North Totals			339	43.65	39.08	372	
⊞ South			470	49.65	45.37	496	
⊞ West			221	37.65	34.70	242	
Grand Total			1111	162.60	148.16	1225	

Figure 2.8: Adding aggregates to pivot tables.

The ability to dynamically add totals and other aggregates helps users roll up information to see the results from a given dimension or set of dimensions. Here, both "boys" and "girls" categories are subtotaled, and each region is totaled as well. Pivot tables are designed to work dynamically, allowing the user to explore and visualize data in whatever way they want. For example, I can remove the "Category" dimension, and the pivot table re-calculates, as shown in Figure 2.9.

Units / Price / Cost		ShipDate				
		1/1/2005			2/1/2005	
Region	Shirt Style	Units	Price	Cost	Units	Price
⊟ East	Tee	25	10.53	9.67	36	
	Golf	27	10.55	9.67	38	
	Polo	29	10.57	9.68	41	
East Totals		81	31.65	29.01	115	
⊟ North	Tee	110	14.53	12.80	121	
	Golf	113	14.55	13.22	124	
	Polo	116	14.57	13.07	127	
North Totals		339	43.65	39.08	372	
⊞ South		470	49.65	45.37	496	
⊞ West		221	37.65	34.70	242	
Grand Total		1111	162.60	148.16	1225	

Figure 2.9: Rolling up data in pivot tables.

I've simplified my view of the data to see only styles, regardless of category. I may have decided that categories weren't a factor in whatever I was analyzing, and by eliminating it, I can view my data and aggregations based on other factors—simply style and region, in this case. This view reveals that the East region sells through significantly fewer units than the North region, a fact that was available to me before but may have been less obvious because it was buried in with so many other numbers.

But here's the problem: Although pivot tables can be useful for users who are already comfortable in Excel, you shouldn't limit this type of BI to just those users. Also, digging in and discovering facts and information *solely* using spreadsheets can become pretty overwhelming. That's why any good midsize BI solution *may* support pivot table functionality, but it will *definitely* support more powerful in-memory analytics, and offer much friendlier and more intuitive user interfaces. These UIs are often presented in a Web browser application, for example. Figure 2.10 shows how a Web interface, powered by a back-end analytics engine, can present data in simpler forms, like a bar chart.

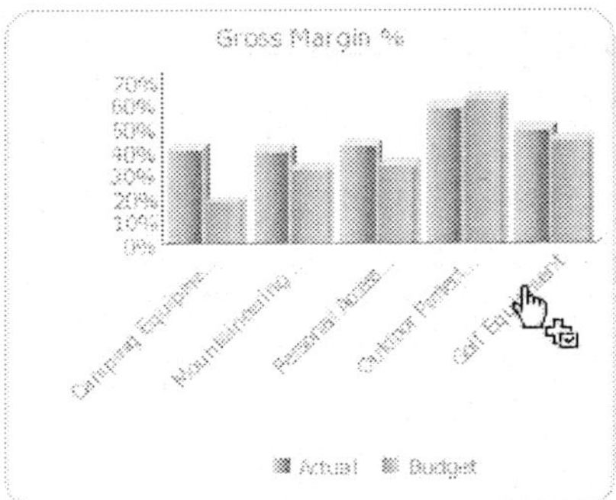

Figure 2.10: Displaying data in chart form.

This chart would be based on live or recently-collected data that was analyzed in memory, meaning it's up to date. But the powerful part about this kind of business intelligence is that *this chart isn't static.* Wondering why golf equipment's margins are the way they are? Just *click on that portion of the chart.* The analytics engine goes to work, pulling together the supporting data that resulted in the first chart and displaying a second "drill-down" chart, shown in Figure 2.11, that shows more detail.

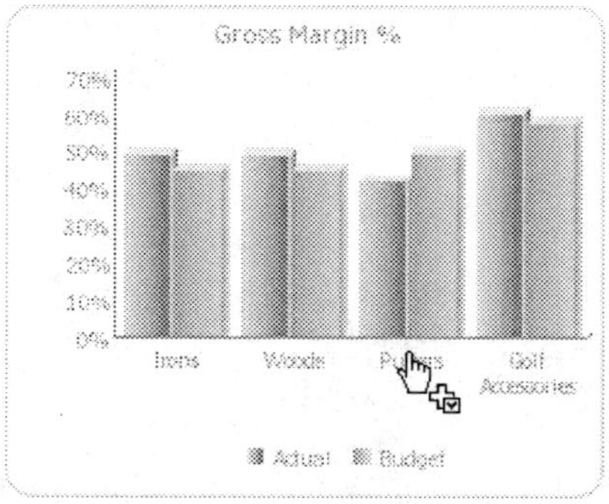

Figure 2.11: Drill-down chart.

Now we can see that putters, as a category, is dragging down margins on golf equipment a bit. Further clicks would drill down even deeper, revealing individual products, manufacturers, and so on, letting us quickly find the answer we were looking for.

But an in-memory analytics engine can do much more. Suppose the first chart, which breaks down sales by product category, isn't answering your questions. Instead, you want to see product sales grouped by store location, or by product vendor, or some other grouping. No problem: Because the analytics are being performed in-memory, you can rearrange and re-group however you like, letting you look at your data in ways that make sense to you, and helping you discover important new relationships and facts.

As you become more aggressive about analyzing your business' data, you'll find that the best suite of tools includes features like pivot tables in a spreadsheet, analytical add-ins to applications like Excel, and Web-based front-ends to powerful in-memory analytics engines. With this combination of tools, more users and managers will have access to better data more quickly. They'll be able to dynamically change what they're looking at to answer exactly the right questions.

Uncovering Root Causes Behind Data

You glance at a dashboard display and see that sales aren't where they should be—what do you do? As I've described earlier, a dashboard and other BI "summary" displays should drive *investigation,* not immediate decision; your response to something alarming—or, for that matter, gratifying—in a dashboard display should be to dig into the root cause, looking either for a problem that needs to be solved or something beneficial that may need to be examined and exploited.

Because BI systems combine data from all over the organization, they excel at providing the details behind the dashboard—for users who know what to ask and how to ask it. In the BI world, asking a question of the system is called a *query,* and the results are typically a report, graph, or other data visualization. The benefit of a BI system is that, unlike more traditional reporting systems built atop transactional databases, BI systems can respond *quickly* with answers—giving users more opportunity to refine, revise, and re-submit their query to find the real answers to their questions.

BI systems typically offer users a variety of ways to submit queries and receive results. Query options may include:

- Natural-language queries. These systems allow users to ask plain-English questions like, "How do sales in the East and North regions compare for last quarter?" The system parses the question and matches key words to data facts, and produces the appropriate report.
- Graphical query builders. These systems use a drag-and-drop interface to present users with available data facts, allowing users to select the facts that interest them, indicate filter criteria, and so forth.
- Query languages. These systems accept queries written in a specific query language, which is often a variant of the industry-standard SQL language. Although this option requires a more technically-proficient user, it does offer a high level of flexibility.

BI systems commonly allow users to save their queries; this makes them available not only for easier querying in the future but also to other, less-skilled users.

And because a good BI system will also make data available through intuitive Web interfaces, your users might not need any particular querying skills in order to drill down, rearrange, and re-group data, and find the answers they need.

The Importance of Modeling the Business

Of course, a BI system can really only work effectively if it accurately models your business. For example, if your business' success focuses on delivering services in a specific period of time, but your BI system's models are focused on facts such as revenue and costs, your BI system won't be as effective in helping you uncover root causes and find new efficiencies and opportunities. Modeling—one of the first steps in a BI implementation, and closely related to data mapping—ensures that your BI system understands what your business looks like, how it works, and how your various pieces of data fit together.

Making Decisions and Taking Action—The Planning Connection

Let's say you've got a great BI system. It contains data from all over your organization, and you've become skilled in submitting questions and reviewing the results. Your decisions are being driven more and more by the information you uncover through your BI system. What next?

The last connection in a BI system is making those decisions a reality. Take a look at Figure 2.12, which illustrates the true purpose of a BI system.

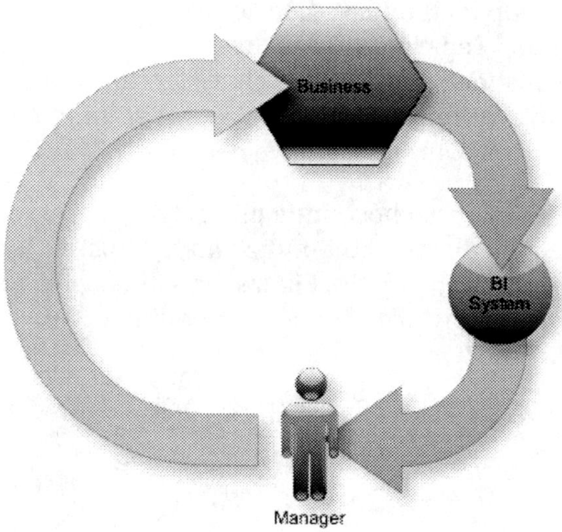

Figure 2.12: The purpose of BI.

The goal of BI is to draw data from your business and use it to inform managers. Those managers' decisions then influence the business, hopefully changing it for the better in some fashion. That change affects the data coming out of the business, which goes into the BI system to influence the next round of decisions.

So the missing piece here is this: *How do managers' decisions get into the business?* In most midsize companies, it's not as simple as managers deciding to run a new ad campaign to boost sales; decisions have consequences such as costs and manpower that have to be accounted for. That means managers' decisions can't always be implemented instantly; instead, they have to impact the company's *planning* process.

Planning: Aligning Resources to Opportunities

Most companies have some kind of planning system and process in place. In many midsize companies, it's simply a collection of spreadsheets—sales forecasts, budgets, capital planning, and so forth. Business managers are constantly looking for better data: What drives the business? Where is the business trending? In many situations, those managers turn to spreadsheets—often hundreds of them, often individually managed by different people throughout the company.

The problem with getting information that way is that it takes a tremendous manual effort to keep it all updated. In addition, learning anything that requires *aggregating* separate spreadsheets is really difficult. What's nice about those spreadsheets, however, is that you can change them: plug different values into different cells, then see what happens. What if we assume a higher sales volume—will we need more staffing budget?

Today's businesses are also becoming more dynamic, and that demands a more interconnected, dynamic planning and forecasting system. That's where your BI system can step in. Figure 2.13 illustrates how the many portions of your company are connected; shouldn't your planning data be just as interconnected?

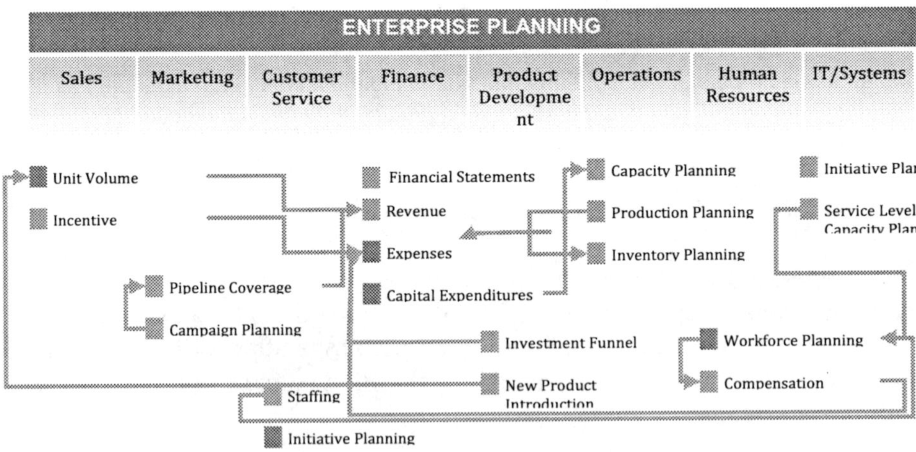

Figure 2.13: Enterprise planning understands the interconnectedness of your company's pieces.

A simple BI system consisting of a data warehouse can provide facts, but it's essentially read-only. Sure, you can pull some of that data into a spreadsheet—perhaps using a pivot table or an add-in of some kind—but then you're right back to dealing with individual spreadsheets. By most definitions, a data warehouse is always read-only: It's a special-purpose *copy* of your data.

More powerful BI systems offer true Online Analytical Processing, or OLAP, including *write-back* capabilities. With write-back, you're actually able to plug new numbers and facts *into the analysis engine*. You're not modifying any of your historical data, but you *are* modifying the way the numbers connect to one another. You're modifying assumptions, and asking "what if," with the power of the entire analytics engine behind you.

Think about how this works. A BI system can connect all of the data in your company. You can dynamically re-group and re-arrange information to answer different questions, all on the fly. That's great for answering questions *about the past*—what has already happened. By being able to reach into the analytics system and make changes, you can start figuring out *what might happen in the future*. You can see sales trends change, see how budgets might be affected, see how you might change your hiring patterns, and so on. You take the latest beliefs, forecasts, and commitments from business managers, and see how they'll affect things like inventory acquisition or customer service volume or production capacity. That lets you make *planning* changes. Suddenly you're no longer *reacting* to your business—you're *driving* it.

What if you write numbers into the OLAP system that predict a downward drift in revenue—something many companies have been facing in recent years? Rather than being caught off-guard and having to merely react to changing markets, you can be in front of the problem, recalibrating expenses and other expenditures so that your company can remain agile and profitable even during a downward swing. It's all based on the capability of your BI analysis engine to take new assumptions, crunch them into your companywide data, and show you what the result looks like.

One of the worst times at most any company is Budget Time, with Quarterly Forecast Time coming in a close second-worst. Sometimes it feels like everyone is just pulling numbers out of thin air, sticking them in spreadsheets, and then hoping for the best. Part of the problem is that everyone's trying to pull data from a dozen different places. They're trying to roll up that data to a profit and loss (P&L) statement level, to create projections on things like cash flow and expenses. Because of the difficulty in doing this manually, a lot of companies don't do it very often—meaning they're running without forecasts, without real *plans.* With a write-back OLAP system, however, all that changes. You're no longer working with decoupled high-level goals and low-level capabilities. Want to see what it would take to double your revenue? Plug the number in, and let the OLAP system *show* you what it would take to reach that number. Want to see what effect a downsizing would have on the bottom line? Plug the number in, and look at the graphs, charts, and reports.

As Figure 2.14 shows, this is where BI really closes the loop for a company. BI is more than just a data warehouse that quickly serves up facts about what *has* happened; a good analysis system can also tell you what *will* happen, when you plug in estimates, forecasts, and what-if scenarios.

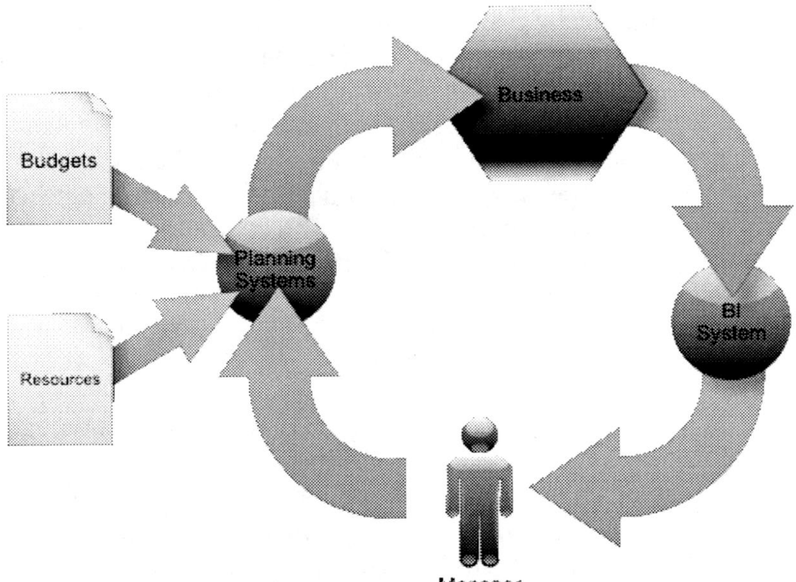

Figure 2.14: Closing the BI loop.

> **Note**
> Write-back isn't a capability supported by all BI systems; if you feel this is an important feature for your business, be sure you take it into account when evaluating different BI solutions.

Closing the loop in this fashion helps create a more automated link between BI, management, and the company's day-to-day operations.

Coming Up Next...

There's a real problem with BI, especially as it relates to midsize companies. That problem is *misinformation.* There are a tremendous number of myths out there that tell you what you can and can't do with BI in a midsize company, and most of those myths are—at best—based on outdated information. In the next chapter, I'll outline the four biggest myths that drive midsize businesses away from BI, or lead them down an implementation path that is unnecessarily expensive and time-consuming. We'll debunk those myths and set the record straight.

Chapter 3: Debunking the Top-Three Myths of Business Intelligence for Midsize Companies

There is definitely a pervasive belief that midsize companies simply can't afford, can't handle, or can't appreciate business intelligence (BI). I even run into executives at midsize companies who absolutely believe that companies the size of theirs don't *need* BI and can't benefit from it. It simply isn't true, and in this chapter, I'm going to play "Mythbusters" and address the most common misinformation regarding BI as it relates to these companies.

Before I do so, however, I want to get one of the most harmful pieces of misinformation out of the way, which is this: *Midsize companies can't benefit from BI.* That's patently false; it's like saying that midsize companies can't benefit from customer relationship management (CRM), accounting software, payroll software, or any other common business tool. Although it's true that many of these technology solutions *began* in extremely large companies, the fact is that *any* company can run itself more effectively and efficiently when these tools are available. It's also true that the CRM solutions (to take one tool as an example) used by giant enterprises are built differently than those used by midsize companies, but that doesn't mean midsize companies don't benefit from their particular versions.

In fact, that's one of the common themes you'll find in this chapter: Midsize companies *can* and *do* benefit from the same business tools and capabilities as giant enterprises; those tools are simply built specifically for the midsize market, and are often easier, less expensive, and faster to adopt and use than the ones designed for giant enterprises. On to the myths.

Myth 1: BI Can Be Disruptive

I find that there's definitely a belief that BI can be disruptive to the business, or at least that the process of *implementing* BI can be disruptive. I've certainly worked with customers who have had dedicated project teams work for many, many months at bringing a BI project to light, and the entire process certainly *was* disruptive to the business in many ways. But those were *gigantic* companies, and many of the problems they dealt with were ones you'd only run into inside a really large enterprise. Let's focus on some of the major aspects of the "disruptive" myth.

Specialized Tools and Training Required

One objection I hear from midsize companies is that they fear they'll need to purchase expensive tools, and undergo extensive training, in order to begin planning and deploying a BI solution. It's easy to see where this "myth" comes from—because I can confirm that it really does happen—*inside huge enterprises.*

But this is hardly true *only* with BI projects. Look at nearly *any* major, enterprise-wide project in a massive company and you'll typically see the need for specialized tools, extensive training, and so on. How many major companies haven't been through an exhausting implementation for accounting software? I remember when one past employer—a 22,000-employee telecommunications division—implemented new financial software and "specialized tools and training" was an understatement. Horror stories abound for CRM and Enterprise Resource Planning (ERP) solutions, as well, and I've seen those projects (in one case I got roped into one). There was a *ton* of specialized tools and training floating around. But again, that was in a *very large* company.

Here's why: Large companies tend to be extremely diversified. They have a lot of things going on, and they have departments and divisions that operate almost as autonomous businesses. Simply figuring out what an accounting system needs to look like or how a CRM solution needs to work requires some very specialized experience. BI is probably an extreme example of this: A truly enterprise-wide BI system wants to touch *every piece of data* the company has so that all that data can play a role in dashboards, reporting, analysis, "what if" scenarios, and so on. Figuring out where all that data lives, who owns it, how it works, and how it will be loaded into a data mart or data warehouse are complicated tasks. The tools used to design the resulting data warehouse are complicated and can be expensive; the tools used to gather information and requirements are equally specialized. But *that's in a very large company.*

So if "specialized tools and training" are a truth for large enterprises, why are things any different for a midsize company? I don't want to come across as suggesting that midsize companies are somehow more generic than larger enterprises, because that isn't true. But the fact is that most midsize companies *do* operate by a more common set of rules and practices than large companies do. That isn't a downside of being a midsize company; it's a *benefit* that most large enterprises often wish they still had. Bear with me for a moment while I explain.

Start by considering payroll. Most midsize companies have fairly basic and common payroll needs: You pay employees, you keep track of paid time off, you account for tax withholdings, and you likely extract some money for benefits. No matter what line of business you're in, payroll tends to look the same—so much so that midsize companies often outsource their payroll entirely or use off-the-shelf software packages that all work the same way. Midsize companies don't often have a massive cadre of HR folks, and so those companies tend to keep their payroll within the realm that can be handled by outsourced companies or by off-the-shelf payroll packages. When you get to be a huge company, however, and you build up a huge HR department, it starts to be easier to do different things with payroll, and so payroll becomes almost a side business within the enterprise—specialized to the enterprise's needs.

What about accounting? Midsize companies tend to follow Generally Acceptable Accounting Practices (GAAP), and there are numerous off-the-shelf software packages that provide perfectly acceptable accounting capabilities. Midsize companies do accounting because they *have* to, not because they especially enjoy it; therefore, they tend to keep their accounting practices within the domain that an off-the-shelf (or lightly customized) accounting package can handle. Sure, midsize companies could start doing complicated things with accounting, and they'd save some money—but they wouldn't save enough to make it worth the extra complication. Large companies, however, can see bigger savings from more complicated and detailed accounting practices, and so those large companies invest in specialized software, specialized processes, and so on.

This same trend continues through just about every business tool: Midsize companies *could* have the same level of complexity as a giant business, but there's not a huge return in doing so. Therefore, midsize companies tend to have back-end systems that all look and work in a very similar way. Those systems, to a degree, drive (or at least influence) the company's business processes and workflows, meaning that most midsize companies look—from a back-end perspective—if not identical, then at least very similar. There's an important truth here that I want to emphasize: *Midsize companies often steer clear of very customized tools and processes because those customizations add complexity that are not justified—to a midsize company—in terms of their added value.*

So how does this affect BI implementations in a midsize company? The reason a massive enterprise needs specialized BI tools, skills, and training is because every single massive enterprise looks entirely different from the others. They've all adopted their own customized accounting, payroll, ERP, CRM, and other practices. Designing a BI infrastructure that utilizes all those data sources will, obviously, be a customized undertaking and will require specialized skills, tools, and training.

A BI solution designed specifically for midsize companies *still requires specialized tools and training*, but because midsize companies look so similar on the back-end, those specialized tools and training come into play *once,* when the BI solution is initially created by its vendor. After that, the same pre-made BI solution will—with perhaps some light customization—be suitable for most midsize companies *simply because* most midsize companies are using the business practices and patterns that were taken as assumptions when the BI solution was created in the first place. In other words, a midsize company can do without the specialized tools and training *simply because* a midsize company will tend to steer clear of the customizations that make the tools and training necessary for an individual BI deployment.

This is an important concept for debunking many of the BI myths in this chapter, so I appreciate you bearing with me through this explanation. Let me try and wrap it up with a non-business analogy, just to make sure I've made this important point.

Consider a NASCAR car. That car requires some expensive, specialized tools, and its driver requires extensive, specialized training. Why? Those tools and training can bring the hundredths-of-a-second advantage that means a win. Most normal drivers—like you and me—don't need a hundredths-of-a-second advantage, so we make do with cars that don't need highly-specialized tools. That means we can very feasibly purchase and own a car *without* having any specialized tools or training. The cars we buy are designed to work in a standardized fashion, even though the cars available to us vary widely in specific features and capabilities.

That's basically why midsize companies *don't* need specialized tools and training to do a BI implementation: Pre-packaged solutions that are *designed for the way most midsize companies already work* have the specialized tools and training built right in.

Long Implementation Times

I was on a BI project in a 40,000-employee company that took *2 years* to complete. No kidding. We spent the first 9 months on requirements gathering, something like 3 months designing reports and data warehouses and such, and a year on the actual production implementation and the inevitable, "oh, can it also do this?" requests from the system's users. Why so long? Well, it was a huge company. Data was everywhere, and we had to reverse-engineer how the company worked in order to build an effective data warehouse. That's kind of a worst-case example, but it serves to illustrate the point that BI *can* take a very long time. In big companies.

I won't belabor the point I made in the previous section any further, except to say that because most midsize companies use similar back-end tools and processes, it doesn't have to take nearly as long to implement a usable, effective BI solution in those companies. Sure, this still isn't a case where you double-click Setup and are done—but it's closer to that than the implementation that giant companies undertake.

In many cases, prepackaged midsize BI solutions are "all in one" packages. You install a single piece of software and you get a place for a data warehouse, Web server, data-loading components, in-memory analytics, reporting, dashboarding, and everything else a BI system needs. You just point it at your existing business data and let it go to work. I've seen implementation times of a few weeks or so, using the existing IT staff. Some companies might be able to get a BI solution up and running in even less time, especially if the solution is initially only addressing a single division or department within the company, simply because that means fewer data connections that have to be made.

Impact on Other Business Processes

I think the first two pieces of this myth pretty much handle this third one. If midsize companies can implement a prepackaged BI solution in a few weeks, using their existing IT staff, and without having to offer extensive training to half the company employees, well, then the BI implementation isn't going to be very disruptive.

Most of the "disruption" in a large-scale BI deployment comes from the discovery and requirements-gathering phases, where BI project team members invade every corner of the company trying to figure out what data lives where, how everything fits together, and so on. The IT staff gets distracted with new hardware, complicated new software, complex data connections, and so forth. With a prepackaged BI solution, there's simply no opportunity for any of that to occur: You don't *have* a months-long discovery and requirements-gathering phase, and the IT staff is usually dealing with a single, simpler piece of integrated software that's much more within the scope of what they're used to dealing with.

Myth 2: BI Requires Specialized Expertise

When I talk to executives in midsize companies, I think the first and loudest objection to BI is the belief that the company is going to have to hire a cadre of specialized (and expensive) consultants to make BI happen.

That's certainly been true in many large companies I've worked with, mainly because their IT staff lacked the specialized skills it takes to plan and implement a data warehouse and other BI elements. Even after the main implementation is over, consultants and trainers usually stay around for several months teaching the company's executives, managers, and other users how to operate the BI solution—and in some cases, I've seen some consultants turn into full-time employees, specializing in BI and in helping the company's users take advantage of the system. But midsize companies don't need to deal with that.

Here's another analogy: Consider two companies that build homes for a living. One company only builds custom homes that cost millions of dollars; the other builds homes in master-planned communities and charges a few hundred thousand dollars. Both companies build great-looking homes, and they use many of the same raw materials. They're also held to the same building codes and other practices, and they both believe in creating a high-quality product.

When the first company begins architecting a new home, a lot of specialized tools are required. Architects have to draw every aspect of the new home, and engineers have to figure out the roof structure, load-bearing capacities, and so on. Because their customers are paying huge sums of money, every job is completely customized, and so the architects and engineers get involved every time. The company's customers all lead wildly different lifestyles: Some want in-home recording studios, others want massive multi-car garages, and so on, so all that customization is really a big part of the business.

The second company's customers don't want to spend millions on a home, though. So the second company offers a few pre-designed floor plans, all of which have been architected and engineered in advance. These aren't cookie-cutter homes; the builder understands which walls are load-bearing, for example, and can do minor customizations that don't affect the pre-engineered structural integrity. This company's customers don't demand a high level of customization—for example, they simply want to pick out their cabinets and countertops. Because these customers don't *want or need* designed-from-scratch homes, they save themselves the expense of an architect and engineer. It's not that those specialized skills were never needed; they were simply only needed *once,* and the results of their work will be used many times, by many different customers.

This "one time engagement" of specialized expertise is what debunks this myth for midsize businesses. Let's look at some specifics.

Complicated Deployment and Implementation

Although a pre-engineered midsize business BI solution isn't exactly as simple as "double click Setup," it is, as I've already mentioned, not a *lot* more complicated than that. At least, it's certainly not as complicated as designing a whole new BI solution entirely from scratch. Deployment and implementation *does* start with double-clicking Setup, often installing a completely integrated, one-piece (or few-piece) solution that contains pre-built BI components such as a data warehouse, reports, and so on. Deployment and implementation is about the same complexity level as deploying and implementing any other server software, such as a database server or messaging server.

The most complex part of the implementation is often the part where you connect the BI system to your other business systems. Typically, prepackaged BI solutions offer "wizards" and other tools to help make this process easier. Prepackaged solutions may even come with built-in "connectors" for common midsize business back-end systems, such as accounting software, ERP systems, CRM systems, and so forth. So although this isn't a "no brainer" part of the BI solution deployment, it's certainly something that the average IT staff can handle—without highly-specialized skills.

Specialized Technology Management Skills

There's also a concern that midsize companies don't have the necessary skills in their existing IT staff to *manage* a BI solution on an ongoing basis. Fortunately, that's not generally a concern. In fact, even some extremely large enterprises get away with very little in the way of specialized technology management skills.

There are two aspects of managing a deployed BI solution. The first is the simple, basic, day-to-day operational stuff: keeping databases tuned for performance, backing up servers, keeping servers properly patched and updated, and so forth. This kind of maintenance is something any normal IT staffer should be able to handle; the most complicated maintenance task—database tuning—is frankly something that can be more or less set up once and then scheduled to run automatically; prepackaged BI solutions typically have this built-in and are advertised as "self-maintaining."

The other aspect of BI solution management is ongoing customization. Big companies change their patterns and practices over time, evolving their business. Some large enterprises make sweeping changes fairly often—and feel the operational pain when *all* their systems, including accounting, ERP, CRM, and so on, need to be updated to reflect the changes in the business model. For those companies, having specialized IT skills on staff is a cost of doing business. Plenty of huge companies, though, make relatively few changes to their basic patterns and practices. They don't *want* to re-architect their accounting records, re-design their CRM solutions, or re-design their BI solution. So those companies, even though they're quite large, often do without specialized BI skills on staff. Midsize companies also tend to avoid the kind of complex changes that require sweeping changes to their back-end systems. Remember, *midsize companies often steer clear of very customized tools and processes because those customizations add complexity that is not justified—to a midsize company—in terms of its added value.*

Large-Company Skills Don't Scale Down

Some midsize companies are fortunate enough to have BI experience on-staff. Typically, it's in the form of an IT staffer who perhaps worked on a BI project in a previous job—often with a large company. Or it might be an executive or manager who used BI solutions in past jobs. The fear, however, is that those large-company skills won't scale down to the midsize business, meaning those existing skills are at best useless, and are at worst dangerous, as they will drive big-company decisions that aren't appropriate for a midsize business.

In my experience, nothing could be further from the truth. I've spent a lot of time writing about prepackaged BI solutions for midsize companies; it would be very easy to assume that "prepackaged" somehow means "different" or "less capable." Not so. Prepackaged BI solutions *are* BI solutions; they work in much the same way that big companies' BI solutions work, and past experience with *any* BI system will make someone more comfortable with *any other* BI system. "Prepackaged" simply means "assembled for you" rather than "build it yourself;" custom-built BI solutions are *custom,* not necessarily better.

So large-company skills don't *need* to "scale down" to midsize businesses; those skills are valuable exactly as they are.

Specialized End-User Skills

Do BI systems require specialized end-user skills? Well...yes and no. This *isn't* a case where I'm going to tell you—again—that midsize companies simply don't need the same skills as big companies. As I outlined in the previous section, big-company BI skills work well in a midsize environment; that suggests that any BI implementation will need the same skills, so someone who *doesn't* have those skills will need training. Right?

Again, yes and no. Let me fall back to the example of accounting systems. In the very earliest days of computerized accounting systems, computerized accounting was only done by big businesses. A *lot* of specialized skills were involved; midsize businesses stuck with pencil-and-paper ledgers. As more and more big businesses used those accounting systems, however, the folks who created those systems found ways to make them easier to use. Eventually, that resulted in prepackaged accounting systems—like today's QuickBooks for small businesses, or Microsoft Dynamics for midsize businesses—that were easier to use and really didn't require a lot of skills specific to the software. Any competent accountant, accompanied by a good instructional book, could figure out the software and use it effectively.

In other words, the investment by big companies made things more accessible to smaller companies. We see that again and again in the IT industry: Big companies were the first to invest in large, mainframe computers in the sixties; today, every company has several PCs. The same has held true for BI: The investment by big companies—in training their end users, in implementing BI, and so on—has made things much more accessible to end users. The first BI implementations relied on complex reports, proprietary analysis interfaces, and so on; today, Web-based dashboards and scorecards are accessible to more users. Add-ins to familiar applications such as Microsoft Excel put BI analysis in the hands of more people, with less training. Big companies started with simple, information-dense charts, like the one shown in Figure 3.1.

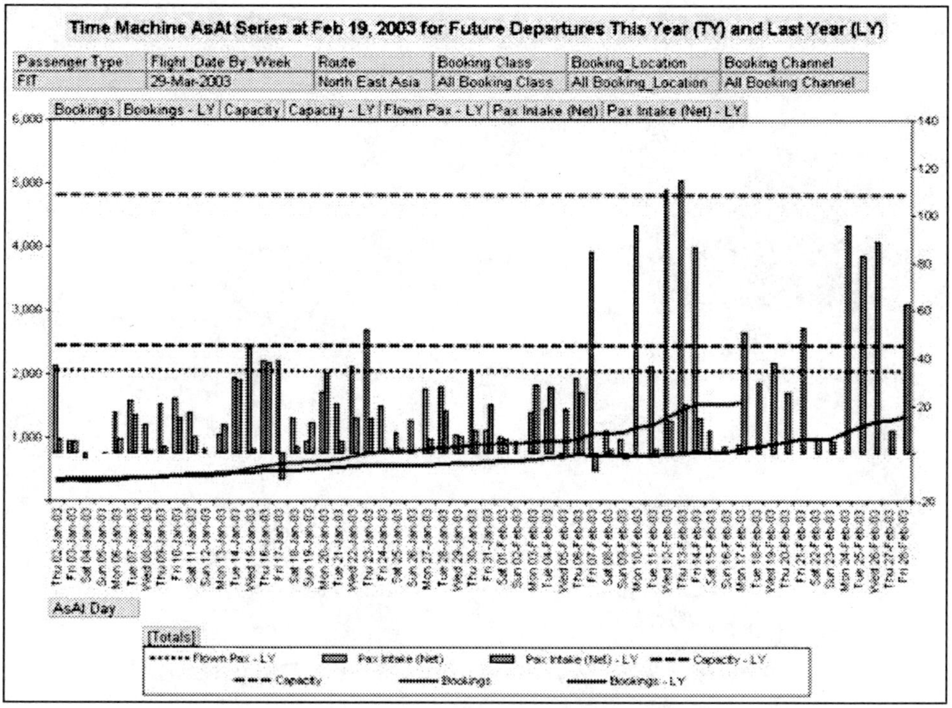

Figure 3.1: Detailed BI chart.

Although this type of chart is still useful today, it took the first BI users a good deal of time and effort—using specialized skills—to pull this information together into this display. Today, BI solutions for both enterprises and midsize companies can automatically produce richer displays, like the sales dashboard shown in Figure 3.2 and the scorecard shown in Figure 3.3.

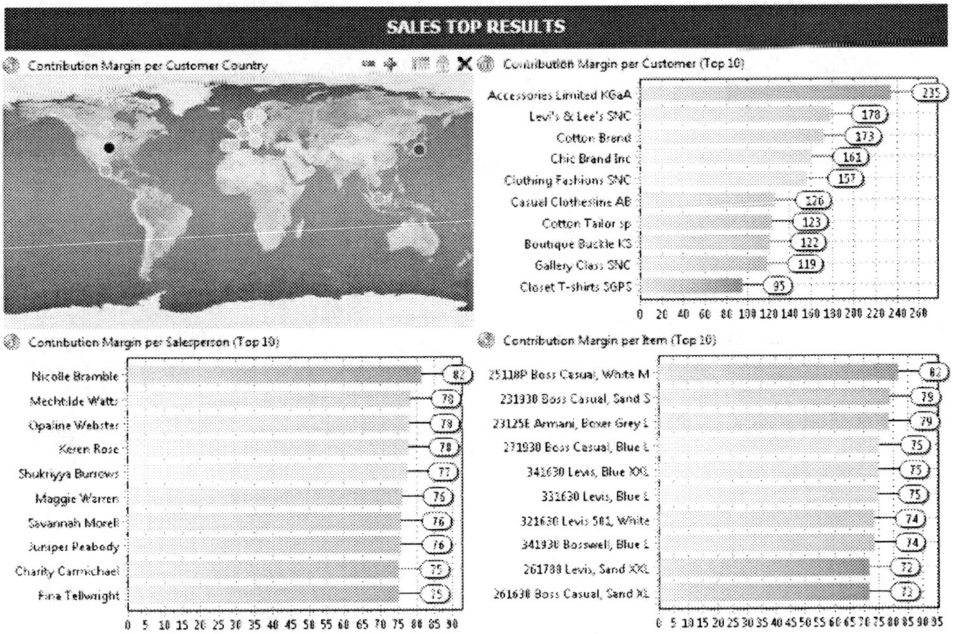

Figure 3.2: Sales dashboard from a BI system.

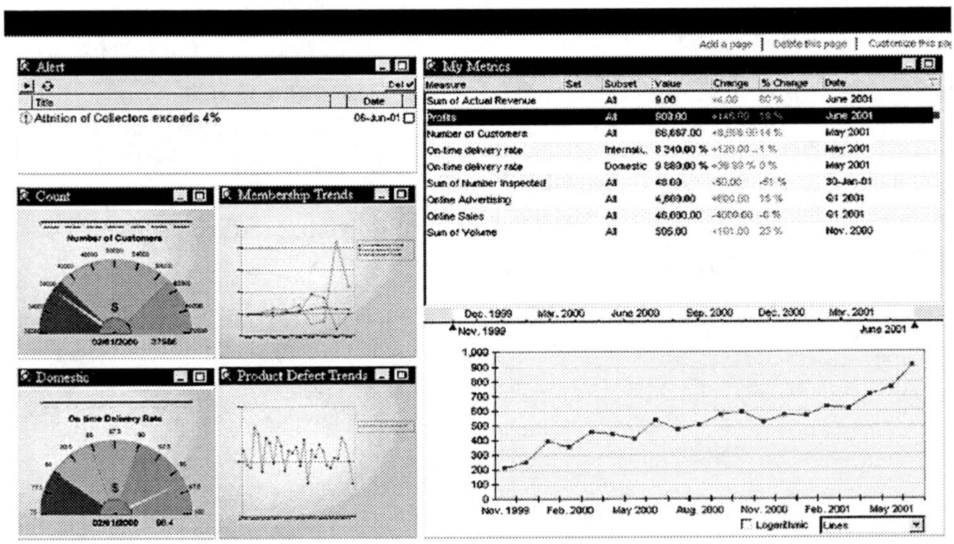

Figure 3.3: Example scorecard from a BI system.

Again, the answer to this myth isn't the same "midsize companies aren't the same as big companies" answer I've provided up to this point; in this case, specialized end-user skills are less needed by *any* BI system—large or small—simply because BI solution developers have made considerable effort to make things more intuitive, automatic, and approachable. We're simply seeing the culmination of more than a decade of continued progress and evolution.

Now, that's not to say that big companies don't *have* end users who have specialized skills. If you're trying to answer entirely new questions using a BI system, then things will certainly go faster if you're experienced using your toolset. But in large part, I think the operative word there is *experience. Any* intelligent businessperson can gain "specialized skills" simply by using their BI toolset for ever-more-complex tasks. And "training" for these end users is also more widely available: Type "BI data analysis" into an online bookstore's search engine, and you'll find hundreds of results.

Myth 3: BI Is Expensive

When I've worked with big companies, there was no question that BI was a pricey proposition. But isn't *everything* in a big company expensive? That's part of the deal of being "big." But you can't use big-company pricing to draw accurate conclusions about BI costs for midsize companies. Your midsize company didn't spend as much on, say, a CRM solution as a gigantic enterprise—why would BI be any different?

Expensive Acquisition Costs

There are two main costs associated with a BI implementation, and one of them is obviously the cost of the BI software. That is what it is, but a midsize company shouldn't expect to pay anything like what a huge enterprise would invest. The vendors that produce these solutions aren't stupid; they know perfectly well that they need to offer different price points for different-sized businesses.

Many of these vendors do so by offering different packages or "editions" of their solutions and toolsets. Just as Microsoft offers an "Ultimate" edition of their Office productivity suite, they assume only power users would purchase that edition. Other editions—"Home & Student," "Professional," and so forth—all provide different price points, and functionality packages, for users with different needs. Today, BI solutions aren't really that different: Your midsize company can get the functionality you need in an "edition" that's much less expensive than what a big enterprise would need and be willing to pay for.

Expensive Consulting Services

The other main cost associated with a BI implementation is the hordes of consultants who will spend months and months researching, planning, implementing, and training before your BI solution is officially online—in a big company, at least; as I've already discussed. Midsize companies can benefit from prepackaged solutions that require little or no consulting services. Sure, you'll probably spend some time speaking with a salesperson and a technical expert, selecting the right feature set for your needs, and making sure that the solution you eventually purchase will do what you expect it to. But in many cases, these prepackaged solutions are *prepackaged* in every sense of the word: You don't *need* to have a bunch of outsiders come in and install it for you.

Notice I said *need.* You certainly *can* have consultants handle your BI implementation, and there are some valid business reasons for doing so, such as an IT staff that simply doesn't have time for the project regardless of whether they have the right skills. But with BI solutions targeted specifically at the midsize market, you'll find that even the consulting expenses—should you choose to use consultants for your implementation—are much lower than what a big company would expect to pay. You're looking at *weeks* of implementation, not *months;* you're dealing with a BI solution that's intended to be installed using fewer specialized skills (meaning less-expensive consultants), in less time, and with less effort.

Don't Like Negotiating with Vendors?

There's another thing I find about midsize companies who are exploring BI options: They hate dickering with salespeople. There's this feeling, nowadays, that every business software purchase is going to be like buying a car. You have to haggle over the basic license pricing, then haggle over the maintenance fees. The salesperson is trying to figure out exactly how much he can squeeze before you kick him out of the conference room, while you're trying to figure out how low you can drive the price before the salesperson gives up and walks away.

Some businesspeople love negotiating, and there will always be salespeople willing to do so. But some BI vendors—especially those targeting the midsize market—realize that a lot of midsize businesspeople don't have the time or inclination for protracted negotiations. They're buying a prepackaged solution, after all; why can't they pay a prepackaged price?

You'll find that *many* BI vendors—again, especially those targeting the midsize market—provide simple, fixed pricing on their midsize business BI solutions. You're just buying a piece of software, not a piece of real estate, so they try to make things easy as well as affordable.

Look for a Trial Version

Some BI vendors make their midsize business solutions available as a free trial download. To me, that says an awful lot about the myths we've explored in this chapter.

A free download means the vendor is pretty sure you'll be able to set the software up and use it quickly (before the trial expires), and that you won't need specialized consulting services or skills in order to have at least basic success with the software.

A free download means you're not expected to have any specialized planning or design tools, either. A free download is also a sign of software that is, or is on its way to becoming, somewhat commoditized. That's not to say that different vendors don't have competitive advantages, but rather to say that they've really gotten the software *prepackaged*, and that obtaining it and using it shouldn't be any more complicated that obtaining and using any other piece of server software.

Plus, a free trial means you can often install and explore the software entirely on your own, at your own pace, and under your own conditions—you don't have to engage in a lengthy sales relationship just to see what the software looks like, what it does, and how it works. I find that midsize companies—especially their IT departments—really love being able to test something on their own, *then* engage the vendor's sales organization when they're ready to proceed.

Unique Advantages of Midsize Business BI

There are a few things that midsize companies can typically do with BI that is actually a distinct advantage, giving midsize companies a real benefit that is more difficult for larger companies to achieve.

In-Memory Analytics

One example is in-memory analysis. As I described in the previous chapter, it's a growing trend to use in-memory analytics because they're fast, can often access more up-to-date (or even real-time) data more readily, and ultimately because they provide faster answers to questions. They allow more rapid exploration of "what if" scenarios because you don't have to wait for reports or other output to be generated.

Big companies can certainly use in-memory analytics, but it can often be expensive for them (remember, *everything* a big company does seems to be expensive). Because their business models deal with *so much* data, they need a *lot* of computing power—especially memory—to make in-memory analytics possible. Some big companies choose to live without it, while others make the investment. Midsize companies often deal with a smaller volume of data, meaning in-memory analytics is more immediately approachable. Computer hardware costs the same no matter how big your company is, and a midsize company will often need much less of it to enable in-memory analytics—a distinct advantage given the benefit in-memory analytics can have on business decisions.

Consistency

Big companies are...well, they're big. They have lots of divisions with different needs, and that can make BI difficult. People wind up looking at different, customized reports, and drawing different conclusions from what they see. Another advantage of midsize companies is that everyone tends to be a little closer to the actual line of business, so everyone tends to be a little more consistent in their needs. Prepackaged BI solutions capitalize on this by making consistent BI output available across the business, in the form of reports, dashboards, and so forth. By getting everyone literally on the same page—something a midsize company can have an easier time doing—you can make more consistent business decisions and get everyone moving in the same direction more easily.

What's Good for Big Business Is Kind of Good for You Too

Hopefully, I've convinced you that BI isn't just for big businesses, and that many of the common perceptions about BI are in fact *mis*perceptions. BI doesn't have to be expensive, doesn't have to involve an army of consultants, and doesn't require your business to slam on the brakes and disrupt itself.

But in using terms like "prepackaged," I may have given you the impression that midsize business BI offerings are somehow less capable or flexible, and

I need to address that. Although BI solutions targeted at midsize companies often do include a subset of the features found in a "big company" BI solution, that's typically because many of the "big company" BI features are specific to a major enterprise environment and those features often *do* require the specialized skills and training midsize businesses are less likely to have or want to acquire.

That does *not* mean that prepackaged, midsize business BI solutions are inflexible, "one size fits all" programs that will force you to change the way you do business in order to "fit" the solution's preconceived model. Modern software is, fortunately, much more flexible than that.

Business Model Flexibility

Every BI vendor knows that they can't force you to remodel your business to fit the vendor's notions of how your business works. If they tried, you'd simply never by their product, and they'd fail. The value of BI can only be realized when the BI solution *maps itself* to your business model, not the other way around; that's why large businesses typically have customized BI solutions built for them.

It's the *similarity* of midsize businesses that make prepackaged BI solutions feasible; nobody expects midsize businesses to be *identical.* BI solutions targeted at the midsize market can be incredibly flexible, and exploring a solution's ability to map to your business model is one of the first things you should do when evaluating solutions for your business.

Advanced Reporting

"Midsize" doesn't mean "dumbed down." Midsize business BI solutions typically include powerful, customizable reporting capabilities. A difference, however, is that these capabilities are typically exposed in a much friendlier, more intuitive way. BI data is often stored using a common metadata model that uses standard business terminology rather than tech-geek-speak, making the BI system more understandable to business users. Midsize BI solutions typically include lots of built-in reports for the most common business needs, and these reports can serve as the basis for additional, customized reports that are shared—through a central repository—across the entire business. As users gain experience, they can usually begin writing their own custom reports, again sharing these through the BI solution with the other users in the company.

Data-Everywhere Accessibility

Big businesses were among the first to invest heavily in mobile workforce technologies, and that investment—like all the others big businesses have made in IT over the decades—pays off for midsize companies. Having created "data everywhere" capabilities for their big-company BI solutions, BI vendors moved quickly to bring the functionality to their midsize business offerings as well.

And *any* business can certainly benefit from "data everywhere." The popularity of mobile devices such as Blackberries and iPhones proves that business people love to be connected to their businesses all the time—and being connected to your BI solution is no different. Midsize business BI solutions can provide robust mobile support, including the ability to access reports, dashboards, and scorecards from mobile Web browsers, dedicated mobile applications, and even through mobile email (see Figure 3.4).

> **Note**
>
> The continuing improvement to mobile device Web browsers can make even more BI data available remotely, because—as I'll discuss next—many BI systems are relying more and more on Web-based interfaces.

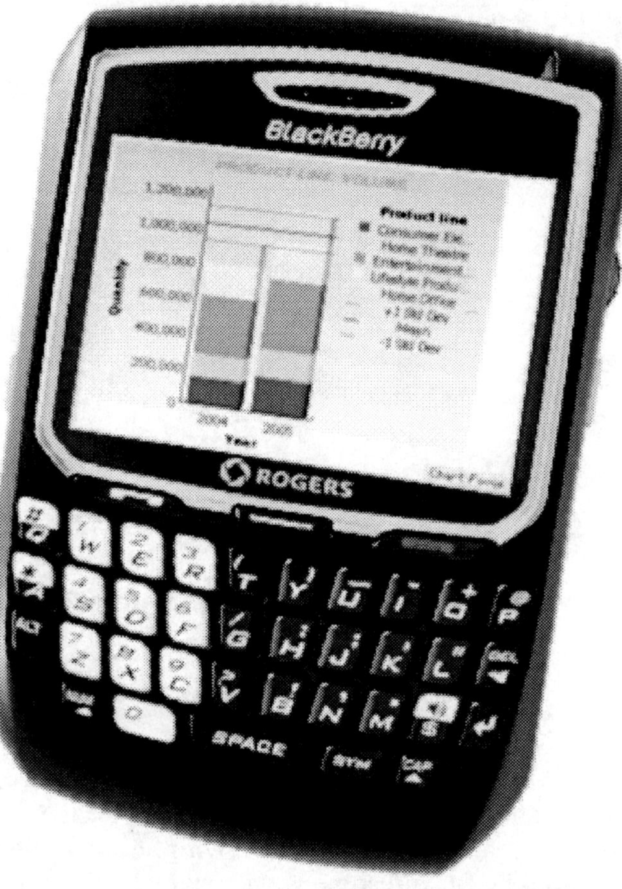

Figure 3.4: BI solution being used from a mobile device.

Mobile devices aren't the only part of "data everywhere" that you should look for in a BI solution. Simply being able to access reports, dashboards, and analysis tools from a home office, from a hotel room while traveling, or in the conference room of a business partner are all valuable usage scenarios. Most midsize business IT solutions accommodate these scenarios by standardizing on powerful, modern Web-based interfaces for much of their reporting and analysis functionality. Figure 3.5 shows an example of a Web-based analysis/reporting interface—something that users could access using any Web browser from any computer in the world.

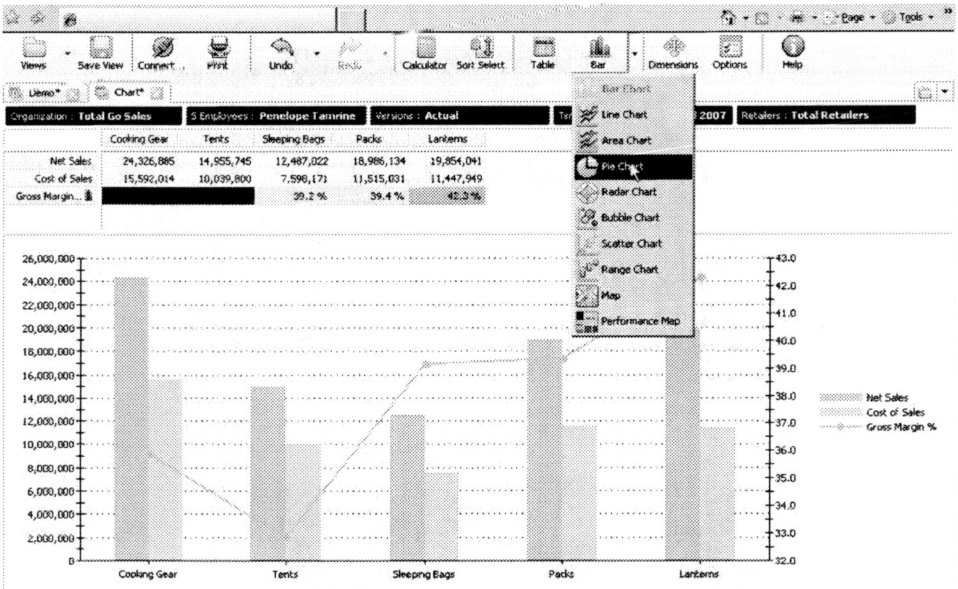

Figure 3.5: Web-based interfaces are accessible from anywhere.

Coming Up Next...

The next and final chapter in this book will outline a roadmap for successfully adding BI to a midsize company. Now that you've learned what BI is, how it works, and what *isn't* true about it, you're ready to see how to bring it into your company. I'll start by reviewing some of the main challenges presented by BI, and some ways in which you can address those challenges in a uniquely midsize company fashion. I'll look at ways to bring BI into the company without turning it into a giant, never-ending implementation—and without breaking the bank. I'll show you how BI *can* be made to work within the scope of your current IT resources, without needing expensive consultants camped out for the next 18 months. Finally, I'll look at how to do BI in a way that complements, rather than disrupts, your business environment. I'll wrap up this book by explaining a few of the things that big businesses deal with when it comes to BI and how *some* of those things aren't good for midsize companies; I'll then show you how to avoid them.

Chapter 4: Successfully Adding Business Intelligence to a Midsize Company

At this point, you should be at least ready to consider a business intelligence (BI) solution for your midsize company—and you may even be outright convinced that it's the right tool for your business. So how do you go about successfully adding a BI solution without disrupting your company? Without breaking the bank? Without having to add staff members with specialties you've never even heard of before? My goal in this chapter is to help answer exactly those questions, with practical advice for bringing BI into *your* midsize company.

Reviewing the Problems with BI

Before we continue, I want to just briefly review and summarize some of the hurdles that BI has traditionally faced. Keep in mind that these hurdles are *potentially* a problem for *any* size company; as I go, I'll outline why big companies choose to accept these downsides, and how midsize companies can avoid them entirely.

Too Complex

BI systems are typically seen as incredibly complex. That's because, to a large part, the underlying business systems and processes that a BI solution must model are also complex. In other words, if you have a huge, complex company, then your BI system is likely to be huge and complex also.

Big companies accept this as a fact of life simply because *everything* in a big company is big and complex. Payroll systems. Accounting systems. Customer relationship management (CRM) systems. Enterprise resource planning (ERP) systems. Asset tracking systems. Heck, just filing an expense report in a big company can be like trying to pass a bill in Congress—*everything* is complicated. To a large degree, big companies can't avoid that complexity. They're subject to a lot of legal scrutiny, for one thing. Microsoft, for example, has to be *very* cautious about how its various business units interact, simply because they're almost continually sued over the results of those interactions.

Midsize companies, however, *can* often avoid some of that complexity—and typically try very hard to do so. Midsize companies don't often have international divisions, multiple lines of business, and mergers and acquisitions to deal with. Midsize companies often use off-the-shelf software or services for things like accounting, customer management, stock management, and so on. That makes midsize companies inherently *less* complicated, and a less-complicated company will have less-complicated options for BI.

The analogy I've used before is that of a house: Building a giant company's BI solution is a lot like building an entire city. No other city is going to be exactly like it, so you wind up with a lot of complex architectural and design elements. You have many different buildings, each of which have a different purpose and must be built for that purpose. You can't use off-the-shelf engineering numbers; you have to work all the math from scratch. Nothing is standardized; everything will be completely custom and there will be a lot of it to deal with. All that customization and complexity makes the building process complex: Permits are complicated (and there are a lot of them) engineering drawings (tons of them) are more complicated, even things like grading plans and landscaping designs will naturally be more complicated. The upside is that you get an entire city in the end; the downside is that it takes a lot of time, money, and effort—and a lot of wasted effort because "custom" also means a lot of wrong turns, backing up, and re-doing things. Another downside is that trying to build it all in one go will mean a lot of backtracking and wasted effort, a lot of re-done work, and a lot of projects that start but simply never get finished.

For me, a midsize company's BI solution is more like a nice house in a master-planned community. There's still a lot of quality in the construction—in fact, the quality can often be higher than the custom mansion because the tradesmen are building several units that are basically the same, and they work out all the glitches early on. The house can still have a lot of semi-custom options and elements, so they're not cookie-cutter (because midsize businesses certainly aren't cookie-cutter), but many of those options are pre-engineered, so they're easier to build and less expensive, and they can be plugged into the base home design without a lot of grief and turmoil.

So are BI systems complex? Yes—if the underlying company is complex. Midsize companies tend to avoid that kind of complexity on their own, however, so a midsize company BI solution is likewise much less complicated.

Too Expensive

Complexity leads to expense, and because BI systems are typically seen as complicated, they're also typically seen as expensive. And they *can* be. Big businesses typically *need* them to be expensive because that expense is what pays for a BI system that can match the underlying complexity of the giant business itself.

Midsize businesses *don't* need to automatically spend a fortune on a BI solution. There's just no need. Prepackaged BI solutions—what I'll call a *software appliance*—can be purchased for a much lower, fixed price than a custom-built, massive-enterprise BI project. Software appliances are entirely self-contained—you may just install a single software application onto a server, or you may even purchase the solution from a value-added reseller who just shows up with a preinstalled machine and plugs it into your network.

With a software appliance or "prepackaged" approach, your costs are also known fully, up front, before you make an acquisition. You can decide then if it's "expensive" or not, and know that the price tag is *all you'll pay*—because the BI solution is prepackaged and self-contained, there's no chance for scope creep, for never-ending implementation projects, or for consultants that have been in one of your offices for longer than half your employees.

Too Disruptive

I've saved my worst-case BI story for this last chapter. I once worked for a company—a large credit card-issuing bank—that decided to implement a major BI project. They were sticking with a specific division of the company, figuring that would be easier and less disruptive than trying to go whole-hog with the entire company all at once. This particular division employed about 12,000 people (although many were part-timers), and had literally millions of customers.

The BI project took *2 years* before it was finally complete and every consultant went home. In the meantime, the division had to bring on another 1000 employees to offset the resources that wound up being tied up almost full-time on the BI project. They spent *forever* figuring out what reports they would need, what dashboards would look like, where data was coming from, where data needed to go, what spreadsheets would contribute and need to be updated—it was nuts. There were seven different data warehouse designs in all, because every time they'd complete one design, they'd realize that it was missing a few things and end up having to start over.

Twenty IT people—including myself, unfortunately—were on this project almost full-time for 2 years. We emerged from it completely unaware of what else was happening in the company. Most of us occupied fairly senior positions in IT, yet we were essentially clueless about the state of our own department because we'd been entirely focused on this BI project for so long.

"Disruptive?" You bet. BI *can* be disruptive. Now, our project was not a shining example of BI even in a very large company, but the point is that BI can get away from you, and that's when it starts being disruptive. Can you imagine a tenth of your IT staff occupied for several months on nothing but a BI project?

The good news is that, with prepackaged BI software *designed* for a midsize company, you'll never *have* to imagine that. There's no "discovery" of business requirements to be done. Many reports and dashboards will come prebuilt, so you won't have to create everything from scratch – you'll be able to start using the solution much more quickly. There are no data warehouses to design, redesign, discard, and design again – in some cases, the solution may even rely more on in-memory analytics than a data warehouse. Much of the design work is done *for* you—all you have to do is install it and configure data connections. It doesn't have to be any more "disruptive" than installing a new messaging server into your infrastructure. Yes, someone will have to spend some time on it—but that time will be measured in days, not months or years.

Adding BI Without Breaking the Bank

There are a few ways you can choose to approach BI, even as a midsize company that plans to use a prepackaged software appliance. You could, for example, go for a companywide BI solution that encompasses everything you do; you could choose to instead focus on a smaller initial project. Your choices in this matter will help define your costs, so what choices can you make that will minimize acquisition and implementation costs, while delivering real, tangible value?

Not Just Reporting!

Before I dive in, I want to re-emphasize one point I've tried to make throughout this book: *BI is not just reporting.* It's easy to see where someone might think BI *was* just fancy reporting because BI tends to focus on dynamic reports, static reports, scorecards, dashboards—all of which are, really, a kind of report. True BI, however, should be the whole, integrated reporting + analysis + planning. Call it *Corporate Performance Management* if you like (some folks do). Gather data. Analyze it in various ways. Make tentative decisions, and see how they affect the data outcome. Make final decisions, and have them modify your planning systems dynamically to make implementing those decisions seamless and easier. Repeat, as shown in Figure 4.1.

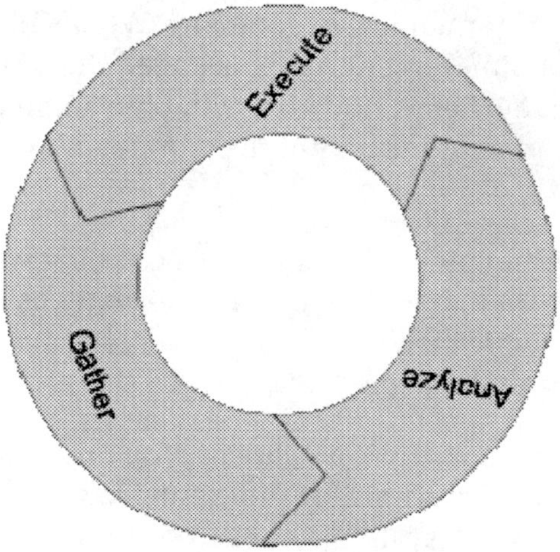

Figure 4.1: The Corporate Performance Management life cycle.

This is what true BI is all about. Not just delivering data in visually-stunning dashboards, but giving you real tools to spot trends and underlying causes, to make decisions, and to *make those decisions happen*.

Think Big, Start Small

In previous chapters, I've mentioned a couple of approaches to data warehouse design. These are both very high-level philosophies, both espoused by very high-level experts in the field. The first is the *top-down* approach, which says that you *have* to start be creating a data warehouse for the entire company, then break that down into smaller data marts for different departments and purposes. The second is the *bottom-up* approach, which more or less states the opposite: That you start with smaller data marts that are very task-specific, then gradually aggregate them into a companywide data warehouse.

Proponents on both sides of the debate have very scathing and witty things to say about their opponents, and I don't propose to have a final answer on the subject in this short book. However, what I will say is this: Philosophy and debate is all very well and good, but the practical realities of the business world sometimes mean you make a compromise. For me, *bottom-up* is that compromise.

Yes, you have to think about your entire company. But implementing a companywide BI solution is obviously more complex than one that just serves a specific department. As we've already seen, complexity = time = expense; a companywide system is desirable but it isn't always the most practical starting point for a midsize company. What's often more practical and approachable is to implement a prepackaged BI software appliance for *one part* of the company. Learn how to use it. Gain some experience with it. Then, gradually, add in another part of the company. And another. And another.

Will a BI solution built in pieces be as good as one that was built from the top-down approach? Well, that depends. My normal answer would be "maybe, maybe not," which isn't very definitive—because it depends a lot on how well the pieces were designed.

Depending on the exact solution you choose, you might wind up with separate, independent BI systems for different divisions—which can be united into a companywide system. Other BI solution vendors may offer a single system that can start out small and grow to include other departments, and eventually the entire company. This is a point to investigate as you begin researching and evaluating specific solutions; neither approach is wrong, but understanding the approach that a vendor has chosen is certainly important.

Start with the Biggest Pain

I once worked for an international retailer. We shipped product to hundreds of stores from three distribution centers worldwide; we also managed a small direct-mail business for customers who didn't have a store in their immediate area. One of the company's biggest frustrations was our shipping costs. We'd negotiate the best deals we could with various carriers, but there was always this sense that shipping was a vast, black pit into which we just poured money. Our executive team was notoriously detail-oriented—they'd re-engineered a number of our business processes and had saved hundreds of thousands of dollars in so doing. But shipping just felt like this untouchable component—it cost what it cost, and all you could do was negotiate lower rates. They were also frustrated with what they perceived as inconsistent delivery times and other details—they couldn't make sense of exactly what their money was getting for them.

It was the first area the company chose for a BI solution. In fact, it was the untouchable, unknowable nature of shipping that made our executives even consider a BI solution. Everything else they'd tried hadn't given them any insight into shipping; maybe BI could prove itself by doing what they couldn't accomplish on their own.

A couple of months later, we had a pretty basic BI solution in place and our executives were already getting answers—surprising ones. Their strategy had been to offer exclusivity to a given carrier in return for deeper discounts, and to send packages that were as close as possible to the carrier's maximum size and weight. Some what-if scenarios in the BI system suggested that even with modest discounts, we'd save money—and get better delivery times—with somewhat smaller packages, and if we used several carriers, selecting them based on the proximity of their hubs to our stores. Within a year, the BI system had paid for itself in shipping savings.

Not long after, we started plugging the BI system into more areas of the company. We could punch in proposed sales figures for a given month and see exactly what we'd probably be spending on shipping, see if the distribution center staff would need more or fewer part-timers to handle the restocking workload, and even predict how long the loss prevention department would need to audit the sales. As I was preparing to leave the company, a single dashboard could even tell you how much loss to expect for a given sales amount (it was surprisingly non-linear—we saw greater losses at very low and very high sales amounts, but less loss for average sales), leading to a whole new set of initiatives to combat that loss. It was exciting.

The lesson is to focus on where you have the *most* pain. Start your BI effort there. There are a few reasons for doing so:

- You have the most to gain. Relatively few and minor insights from a BI solution can provide maximal effect in your most-painful area of business.
- You're probably already focusing on this area of your business, so the BI system won't be seen as a "distraction," it will be seen as a tool to assist your already-heavy focus.
- It offers the highest incentive. People in a well-run area of your business are less likely to welcome a BI initiative with enthusiasm because they'll feel they already have everything under control. By showing them how another part of the business benefits, you'll have an easier time "selling" them on BI later.

So how do you quantify "pain" in an area of your business?

- Look for areas where major decisions are being made primarily from institutional knowledge—that is, by gut instinct from experienced managers.
- Look for areas that are awash in raw data but have relatively little *refined* data.
- Look for areas where managers are already accustomed to using spreadsheets to crunch numbers and make decisions—those are the folks who will benefit first from a BI solution.
- Look for areas where you sense that trends and patterns *must* exist but where those trends and patterns never seem to materialize.
- Look for areas where old assumptions are still driving critical decisions—a BI system can help revise those assumptions using real-world data.

Select the area of your business that has the *most* of these factors, and you'll be giving a BI solution the best opportunity to help you improve the business.

Add Capabilities as Needed

Like most appliances, software appliances are often modular in nature. Need an icemaker for your freezer? Want to add the ice cream maker attachment to your kitchen stand mixer? Want to add a new reporting module to your BI solution? No problem: Extensible appliances are nothing new, and they give you the ability to buy just what you need, just when you need it, so you're saving money and effort.

Every BI solution vendor will obviously follow different patterns for their tool set, but in general, the capabilities break into two halves:

- **Analysis and visualization.** This includes the ability to pull in raw data and generate analyses and visual representations. In-memory analytics may be offered, which helps you generate faster analysis results from real-time, operational data stores. This is where you'll often get visualizations like the dashboard shown in Figure 4.2.

Figure 4.2: Using a dashboard from a BI analysis package.

- **Reporting and ad-hoc querying.** This often includes self-service reporting interfaces, giving you the ability to create custom reports. The benefit here is that the reports are all drawn from a single data source—the data warehouse—and so every user will be working with consistent data. Reports are often columnar, although you'll also be able to design more complex ones. Figure 4.3 shows a sample BI report. The benefit here is that the BI solution provides a toolset that lets users create *their own* reports rather than just using prepackaged ones. This feature helps each user get the exact answers they need at that exact moment.

Budget vs. Actual Sales By Retailer

Camping Equipment

(US Dollars in Thousands)		2004 Budget	2004 Actual	Variance	% Var	2005 Budget	2005 Actual	Variance	% Var	2006 Budget	2006 Actual	Variance	% Var
Americas	Eyewear Store	-	-	-	-	-	-	-	-	-	-	-	-
	Outdoors Shop	21,167	23,503	2,336	11.04%	24,208	24,902	694	2.87%	25,898	43,809	17,911	69.16%
	Equipment Rental Store	3,038	2,149	-889	-29.27%	2,213	3,574	1,361	61.48%	3,717	2,455	-1,262	-33.96%
	Warehouse Store	19,875	20,655	780	3.92%	21,274	23,382	2,108	9.91%	24,317	20,833	-3,485	-14.33%
	Direct Marketing	4,025	3,458	-568	-14.10%	3,561	4,735	1,174	32.97%	4,925	3,036	-1,888	-38.34%
	Department Store	26,195	31,898	5,703	21.77%	32,855	30,818	-2,037	-6.20%	32,050	32,922	872	2.72%
	Golf Shop	-	-	-	-	-	-	-	-	-	-	-	-
	Sports Store	29,785	22,598	-7,187	-24.13%	23,276	35,041	11,765	50.55%	36,443	46,802	10,359	28.43%
	Retailers (total)	**104,085**	**104,260**	**175**	**-30.76%**	**107,388**	**122,453**	**15,065**	**151.57%**	**127,351**	**149,857**	**22,506**	**13.67%**
	% of Total	**30.40%**	**31.31%**			**31.31%**	**30.40%**			**30.40%**	**29.95%**		-
Asia Pacific	Eyewear Store	-	-	-	-	-	-	-	-	-	-	-	-
	Outdoors Shop	26,835	28,860	2,025	7.55%	29,726	31,570	1,845	6.21%	32,833	45,696	12,863	39.18%
	Equipment Rental Store	1,856	1,614	-241	-13.00%	1,663	2,183	520	31.28%	2,270	2,698	428	18.84%
	Warehouse Store	12,234	12,776	542	4.43%	13,159	14,393	1,234	9.38%	14,969	17,678	2,710	18.10%
	Direct Marketing	5,775	4,614	-1,161	-20.10%	4,753	6,794	2,041	42.95%	7,066	7,956	890	12.59%
	Department Store	25,731	21,963	-3,769	-14.65%	22,622	30,272	7,651	33.82%	31,483	36,332	4,849	15.40%
	Golf Shop	-	-	-	-	-	-	-	-	-	-	-	-
	Sports Store	18,867	17,939	-928	-4.92%	18,477	22,197	3,720	20.13%	23,085	25,325	2,240	9.70%
	Retailers (total)	**91,298**	**87,767**	**-3,532**	**-40.69%**	**90,400**	**107,410**	**17,010**	**143.77%**	**111,706**	**135,685**	**23,979**	**113.81%**
	% of Total	**26.67%**	**26.36%**			**26.36%**	**26.67%**			**26.67%**	**27.12%**		

Figure 4.3: An example BI report.

You may find other modules, too, such as modules that support Microsoft Excel connectivity, which can be a boon for spreadsheet jockeys in your company who are comfortable using Excel for data analysis.

Ensure that You Have a Growth Path

Whatever you do, make *sure* you have a plan in the back of your head to extend your BI solution to other areas of your company. Discuss those possibilities with each BI solution vendor that you speak with and let them help guide you toward a solution that can accommodate those growth plans. Your plans might change, but having a solution in place with the right flexibility will help ensure that you can get where you want to go.

Seek Out Fixed-Price Solutions

I've been involved in a number of high-priced software acquisitions, and I *hate* negotiating with salespeople. Frankly, I'm not very good at it. I never know if their first price is really the price or if they *expect* me to haggle them down. If they do, I don't ever understand why they just couldn't bring me that price in the first place.

When it comes to prepackaged software solutions, however, you're rarely going to find yourself having to negotiate—and you shouldn't have to with a midsize company BI solution. Solutions like this are a known quantity for the vendor; in serving a midsize business audience, good vendors will focus on standardized pricing.

While on the subject of price, however, I do have a caution to offer: *Don't* seek out a do-it-yourself solution. Even if you've got the world's best data warehousing and BI expert on your staff, a roll-your-own BI solution is nearly *always* going to cost you more time and money in the long run than a packaged solution. I've seen just a couple of midsize companies go the build-our-own route, and after a couple of years of sustained effort, they finally gave up and just bought something. When they did, they got a system that did *more* than their homegrown system, had it up and running in a few days, and spent less on software maintenance than they'd been paying the two staffers who had developed and were maintaining, their in-house "solution." One of those companies' decision was driven purely by staffing—their BI expert quit, leaving them without anyone who understood how the BI system had been built or how it worked under the hood. They had *no choice* but to either hire someone else—and salaries in that specialization had gone way up at that time—or buy a prepackaged solution.

Adding BI with Your Current IT Team

I've said before that midsize businesses can and should implement a right-sized BI solution *without* needing expensive consulting services or specialized new hires. Here are some tips for doing so.

Software Appliances

Look for a BI solution that could accurately be described as a *software appliance.* I've used that term before, as a synonym for *prepackaged;* what exactly does it mean? Consider your corporate firewall. There are a couple of ways you can go when choosing a firewall—a dedicated firewall software package, which often runs on an OS such as Windows or Linux, or a powerful hardware firewall, like those from Cisco. These all involve a fairly complex level of configuration, and you'll probably need an expert to at least set the thing up initially, if not to help maintain it on an ongoing basis. Configuration often requires something akin to programming, which definitely requires specialized expertise. In some cases, you may need other pre-requisites in order to make the firewall solution work, such as an authentication server, logging server, and other elements.

The point here is that you have to buy the software and/or components, hook it into your infrastructure, redesign your infrastructure to accommodate it, then fully program the thing with all the settings it needs to operate and that you need to fit your business needs. Alternatively, you buy a firewall appliance. This is a box that plugs into AC power and into your network. Configuration is often much simpler, typically Web-based, and usually wizard-driven. You have fewer options to worry about, fewer things you could potentially do wrong, and you likely won't need an expert to get the thing working properly. Many can even self-discover certain information about your network so that you don't have to configure those things manually.

I'm using firewalls as an example because they're a common network element that can be had both as a "complex" solution and as a prepackaged, preconfigured appliance. For midsize companies, I recommend the latter approach for a BI solution: Find an "appliance." It might not be a literal box you plug in, but it shouldn't be far from that. It should:

- Come largely pre-built and pre-configured
- Come with whatever pre-requisites it needs—you shouldn't have to buy or install anything else to make it work
- Require minimal configuration to adapt to your environment, and that configuration should be driven by wizards and other tools to help make the configuration easier and more orderly

That's how you get a BI solution that your *current IT team* can deploy. In fact, look for a BI solution that's available as a free trial. If the vendor is so confident of their ease-of-deployment that they feel they can offer a do-it-yourself trial—with perhaps a short PDF "evaluator's guide" to walk you through setup—then you've probably found something that can accurately be described as a "software appliance."

Modular Appliances

As I've written earlier in this chapter, your ideal BI solution will often come broken into modules so that you can buy just the bits you want. Another approach is to offer "editions," where successively higher editions offer greater and greater functionality. I prefer the modular approach because with it you can buy just what you need; with "editions" I always find myself stuck with the next-higher edition than I really want just because of one or two "must-have" features.

Spend some time working with trial versions, and/or speaking with the solution vendor, to understand exactly what capabilities each module or edition offers. Even if you're focused on saving money, don't forgo critical functionality—doing so could spell failure for your BI project, and many companies don't like to give something a second chance if the first attempt fails. In other words, don't buy *more* than you need, but certainly don't buy *less,* either.

Ensure that You Have a Growth Path

As I've written before, you need to make sure you're buying a solution that can grow with you—and you need to understand *how* that growth will physically happen. Here's an example: Years ago, back when Microsoft Exchange Server was still new, Microsoft released a "Standard" edition and an "Enterprise" edition. The primary difference between the two was their storage capacity: The Standard edition had an arbitrary limit on how much it could store, while the Enterprise edition supported the maximum Windows disk size at the time.

The problem occurred when people on the Standard edition ran out of room and decided to upgrade. There basically *was* no upgrade path; you had to buy Enterprise, install it, and then migrate all of your mailboxes to it—and then decommission the old server, typically without receiving any credit or refund for the cost of *its* license.

Make sure your BI solution doesn't stick you in the same boat. With a modular product, you should be able to just add on new modules as needed. If you're using a product that's built around feature "editions," make sure you have a way to grow to the next edition *without* having to start form scratch. That might involve paying an upgrade fee and entering a new license code, for example, rather than installing a whole new product and migrating your data, reports, configuration, and other work.

Midsize Manageability

How will your existing IT team *manage* your new BI solution? A BI solution—even one that is prepackaged as a software appliance—still has maintenance and monitoring requirements. Someone needs to make sure it's up and running, and that it's running in a healthy condition. If something goes wrong—like it gets low on disk space—you'll want to know *before* that becomes a real problem. Will the BI solution need periodic database maintenance? Who will perform that? How hard is it to do?

The more pre-packaged and appliance-like a solution is, the more I would expect it to do these things largely on its own. Sure, you'll still have to do OS-level maintenance—patches and so forth—but your IT staff is well-equipped to do that kind of basic upkeep. The solution might maintain its own database, and might even take care of its own monitoring, perhaps by sending someone an email if a problem condition is detected. Some solutions might offer options for sending notifications to an operations monitoring console, although the more pre-packaged the solution is, the less I might expect that option. I would certainly expect an appliance-style solution to provide its own monitoring, often through an administrative console.

Ideally, the entire product should be easy to administer, perhaps through a Web-based console. Figure 4.4 shows an example of how easy it should be to add new data connections to the data warehouse, for example: A nice, well laid-out Web console can make otherwise-complicated tasks easier for your existing IT staff to understand and accomplish.

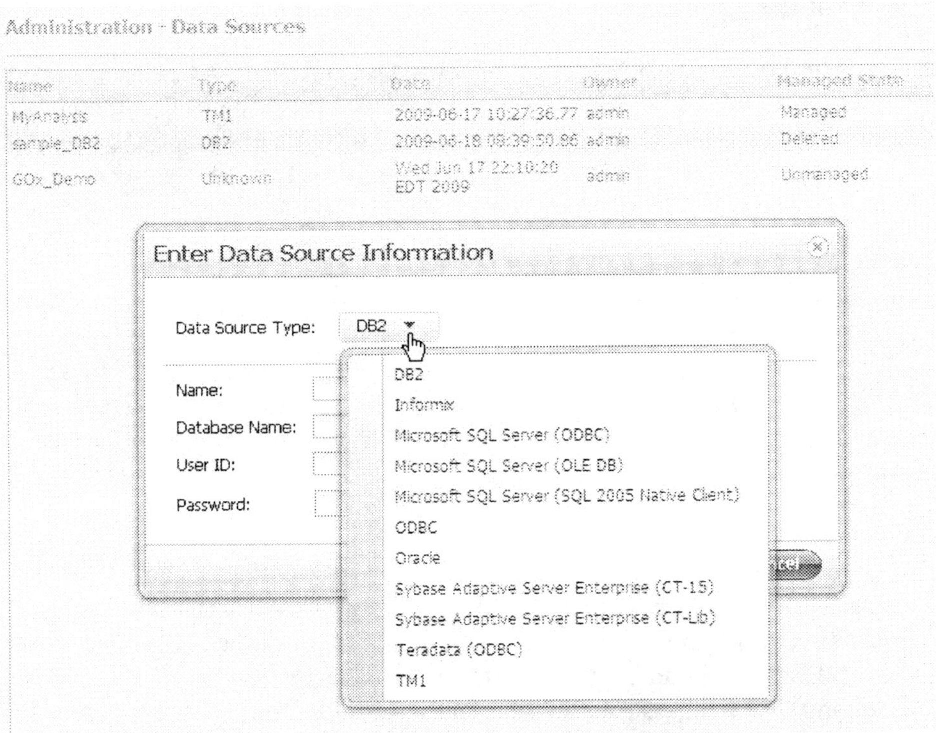

Figure 4.4: Administration through a simple Web-based console.

If a BI solution uses an external database—maybe it requires you to provide a Microsoft SQL Server, or Oracle server, for example—I might expect to handle the maintenance of that database myself. However, such a solution would *not* be an "appliance" in my opinion; appliances don't require you to "Bring Your Own Database."

Adding BI without Disrupting Business

So you *can* deploy a midsize BI solution with your existing IT staff; can you do so without distracting and disrupting your day-to-day business? The trick in doing so is to require a minimum of cost, specialized expertise, and specialized software, as well as a minimum of training. Let's see how a properly-designed midsize BI system might accomplish that.

Low Startup Cost

I've already explained how a fixed-price BI solution is desirable, and how a modular solution design lets you buy just the pieces you need. Those factors, combined with the ability of your existing IT staff to deploy and maintain the BI solution, leads to a lower startup cost. A low startup cost helps contribute to a less-disruptive BI implementation project; fewer eyes will be nervously watching every step, giving the project an opportunity to be completed and put into use without anyone panicking about the mounting expenses.

Besides, high-priced systems in midsize companies almost always lead to agonizing, sometimes antagonistic battles about whether the company should even pursue the solution. That sort of thing puts everyone off-track, disrupting managerial relationships and distracting people from day-to-day business. A lower startup cost lets everyone have a better discussion about the project's merits without quite as many eyes bugging out over an exorbitant price tag.

Finally, a *fixed price* solution makes *everyone* involved with the company's financials feel better. Start talking about consultants and services and fees, and everyone gets nervous because they just *know* the bill will keep going up and up and up. An appliance-style solution with a fixed, known-up-front cost is a known quantity that can be considered and accepted; knowing that the price won't continue to climb is another way to prevent disruption.

Minimal Specialized Expertise

A prepackaged, appliance-style solution avoids that by simply not requiring very much in the way of specialized expertise. The expertise is *built-in* to the product, ready to be used; you don't need much more to get it installed and working.

Self-Service via Web Consoles

Users should be able to access much of the BI solution's analysis functionality through simple-to-use, self-service consoles. Web consoles are more easily accessible from a variety of devices and OSs, and give your users a single, consistent "knowledge portal" through which they can access a range of information. Web consoles can offer rich functionality, too—Figure 4.5 shows an example of a complex analysis that a user can conduct through a Web interface.

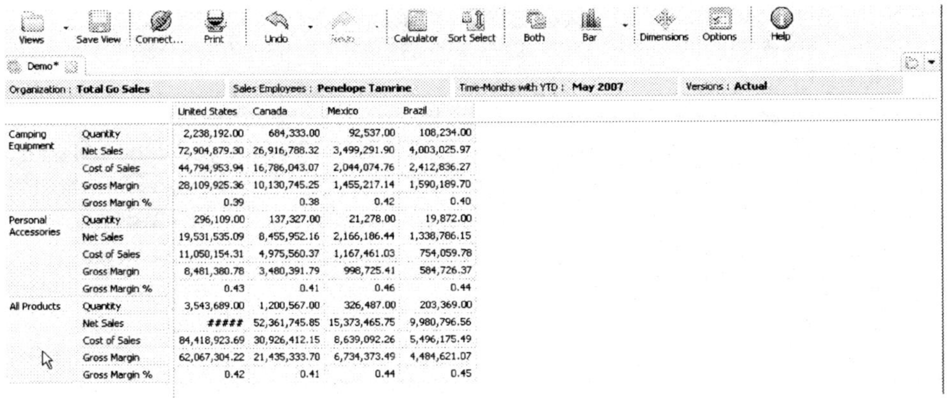

		United States	Canada	Mexico	Brazil
Camping Equipment	Quantity	2,238,192.00	684,333.00	92,537.00	108,234.00
	Net Sales	72,904,879.30	26,916,788.32	3,499,291.90	4,003,025.97
	Cost of Sales	44,794,953.94	16,786,043.07	2,044,074.76	2,412,836.27
	Gross Margin	28,109,925.36	10,130,745.25	1,455,217.14	1,590,189.70
	Gross Margin %	0.39	0.38	0.42	0.40
Personal Accessories	Quantity	296,109.00	137,327.00	21,278.00	19,872.00
	Net Sales	19,531,535.09	8,455,952.16	2,166,186.44	1,338,786.15
	Cost of Sales	11,050,154.31	4,975,560.37	1,167,461.03	754,059.78
	Gross Margin	8,481,380.78	3,480,391.79	998,725.41	584,726.37
	Gross Margin %	0.43	0.41	0.46	0.44
All Products	Quantity	3,543,689.00	1,200,567.00	326,487.00	203,369.00
	Net Sales	#####	52,361,745.85	15,373,465.75	9,980,796.56
	Cost of Sales	84,418,923.69	30,926,412.15	8,639,092.26	5,496,175.49
	Gross Margin	62,067,304.22	21,435,333.70	6,734,373.49	4,484,621.07
	Gross Margin %	0.42	0.41	0.44	0.45

Figure 4.5: Easy-to-use Web interfaces minimize disruption.

Web interfaces also mean you *don't have to deploy software to everyone's desktop.* That's a huge thing, given how disruptive software deployments can be—combined with the training and pain almost always associated with new software.

Intuitive User Interfaces

Speaking of training, you can minimize the learning curve for your users by selecting software that's intelligently and intuitively designed. For example, Figure 4.6 shows an excerpt from a Web dashboard. Someone looking at this might wonder, "Why are margins in golf equipment the way they are?" A well-designed user interface will allow users to drill down and answer that question by doing what the user will tend to do naturally: Click on the "Golf Equipment" item.

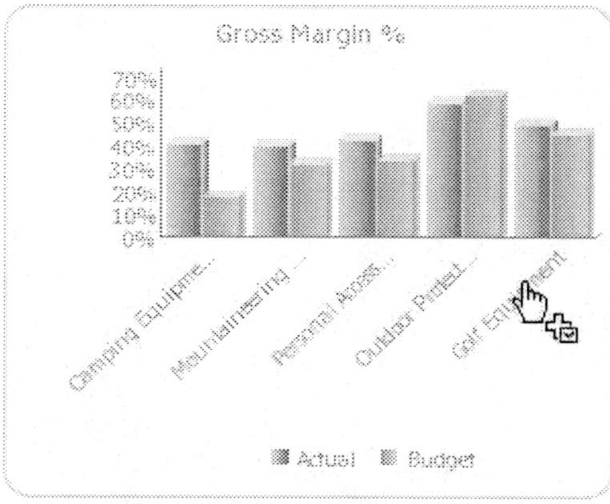

Figure 4.6: Starting with a well-designed user interface...

If clicking leads the user to another layer of data, they'll continue to explore. Figure 4.7 shows where the first click took them—to a breakdown of different types of golf equipment—an intelligent next step in a line of free-form investigation. Users don't need to know any database query languages, or even understand that they're using a BI system or a data warehouse. They're just exploring company data to look for patterns, trends, and underlying reasons.

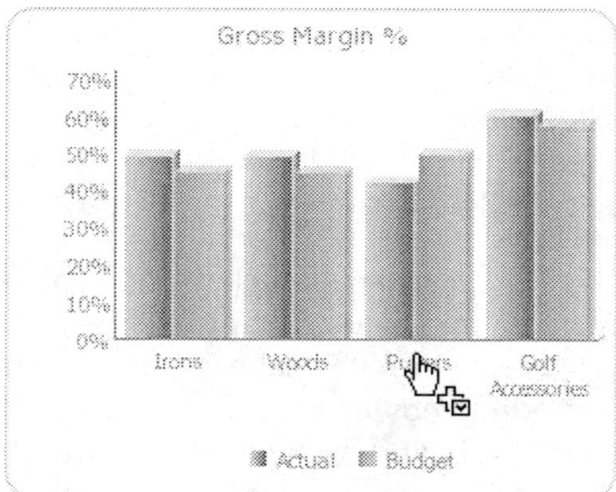

Figure 4.7: ...allows users to drill ever-deeper into the data...

By continuing to support additional layers of drill-down, this user interface allows a user to continue looking for the root cause of the problem. Here, it appears that putters are the lowest-margin item. Should the company just discontinue putters? Another click—and another drill-down—is warranted.

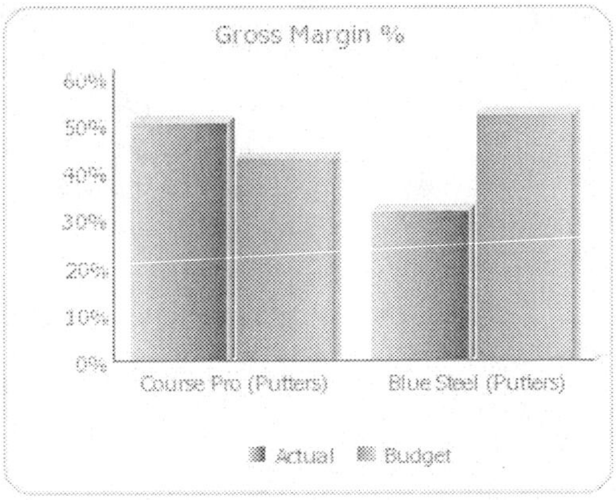

Figure 4.8: ...until they find the root cause.

As shown in Figure 4.8, it's one brand of putters that's actually dragging down the category. Now, this user can make a business decision based on *facts,* not on gut instinct or bad assumptions. It took longer for me to write this narrative than it would have taken a real user to actually reach this conclusion—all without *any* format business analytics training. That's the value of a well-designed user interface: immediate results, less disruption to the business.

Keep Your Existing Tools

Users who are comfortable with an existing toolset should be able to use them, if possible. As I've mentioned before, many midsize business users are incredibly proficient with Excel spreadsheets—so why not let them continue using a tool that they're skilled with?

If you have some of these "spreadsheet jockeys" in your company, having a BI system that supports them—by delivering BI data *to* Excel, and allowing them to work with it there—is a must-have feature. Although Excel isn't for everyone, it can be incredibly powerful and empowering for someone who has invested the time to use it, and because they're working in a comfortable, familiar environment, you'll find that these "spreadsheet jockeys" will be amongst the first to grasp the power of BI, and to start making a return on your BI investment.

What's Good for Big Business...Isn't Necessarily Good for You

Big businesses can often afford a big sacrifice if they see a big benefit. Midsize companies, however, have to be a lot more careful. Disrupting a dozen employees for 3 months is hardly even noticeable for a giant company; for a midsize company, however, it can distract vital assets for an unforgivably long time.

That's why BI isn't "one size fits all." What works for an enormous enterprise—months of fact-finding, specialized consultants to build data warehouse designs, lengthy implementations and even lengthier training—just doesn't fit a midsize business model. What *can* work for a midsize business, however, are smaller-sized BI solutions that—although they would never work for a huge company—fit nicely into a midsize company's needs, available resources, and available time.

I've used the word *prepackaged* a lot in this book, and in the beginning I worried about using that exact term. *Prepackaged,* to many people, seems like it might mean lower-quality, like a prepackaged, frozen dinner. That's not what I mean at all. In the technology industry, prepackaged software products power most aspects of most midsize companies, from financial management to customer relationship management. Even the office productivity suite you're using—a word processor, a spreadsheet, and perhaps a small database application—are prepackaged. Prepackaged does *not* mean lower quality or less powerful. For me, *prepackaged* means an easy-to-install, easy-to-learn piece of software that handles 90% or more of the functions that 90% of business users require. It means *no lengthy, expensive customizations required.* In the world of BI solutions, prepackaged solutions designed to fit a midsize company are the *best* way to implement BI. Frankly, even huge enterprises would use prepackaged BI solutions if they had the option—they'd save a fortune and a lot of time. This is an advantage that midsize companies can capitalize on: The ability to use a prepackaged BI solution brings you big-company power without the big-company price or time commitment.

We've covered a lot of ground in a short space. In the first chapter, I explained what "business intelligence" really is, and why you might want it in your company. In the second chapter, I explained the ways in which BI becomes a reality—starting with how you assemble data and analyze it to make better decisions for your company. The third chapter was a chance to play "mythbusters," debunking concepts like the "fact" that BI can disrupt your business or that BI is only for large companies. In this chapter, I offered some tips and practices for bringing BI into your company in a productive, practical manner.

I hope you've found this information to be useful, and I hope that BI is on your radar not only as a useful set of technologies but as an obtainable, practical idea for helping you and your company make better decisions about your future. Thanks for reading.